Breaking Through Myth to Reality

A Future-Proof View of Fiber Optic Inspection and Cleaning

Edward J. Forrest, Jr.
RMS(RaceMarketingServices)
est:1974

Bringing Ideas Together ™

www.fiberopticprecisioncleaning.com

CONTENTS:

Pages	Description
1-11	Introduction, Background, History of Sciences
12-24	The Myth of Two Dimensions
25-36	Myths of Test and Inspection
37-54	Why Does it Matter. What is the Proof?
55-64	The Role of Standards. Is there an alternative?
65-77	Myths of Cleaning and How to Improve and Improvise
78-91	First Time Cleaning. The Science and Proof.
92-109	Proper Process Selection for Precision Cleaning and Precision Inspection
110-115	Conclusions: Future Proof to Best Practices
116-118	References/Acknowledgements
119-121	Quiz and Answers. Additional resources.

Dedication:

This work is dedicated to those of you who have helped me understand this amazing science. There is an acknowledgement of persons at the back of this book. As well, there are those who can't be mentioned, for reasons we understand.

I also thank my wife, Lanet, who has been an essential part of this wonderful adventure.

Thank you all.

Marietta, GA. USA
January, 2018

Introduction:

In 1990, just as global society and science evolved new standards for environmental safety, I started work in a new career in that field. The concern was "CFC's" and it seemed that any thing we used or manufactured was affected. In the early 1990's, about 100 developed nations signed Clear Air Legislation and the change from solvents, that had industrialized the planet for centuries, began. Along with many others, I was going to get a 'chemistry lesson' that changed my life and career path!

Just about everything from metal cleaning to motors, dry cleaning to how we washed the windows, and, electronics that had been solvent-cleaned to remove soldering flux residues also evolved. The latter was my focus until about 1998 when I received a phone call from Hal Kievlan: *"Ed, do you have anything to clean fiber optics?"*.

Little did I know then what I know now! I would chide myself that the only thing I knew about fiber optics was the lamp I almost purchased from Spencer's Gifts at the local mall! After several visits to a production line in South Carolina where fiber optic electronics were produced, I began to realize, humbly with concern: *"They are cleaning these surfaces all wrong."*

There are many facets to fiber optics. Some are fanciful lamps and Christmas trees and others incredibly intricate and sophisticated devices that have been part of my life since 1998, and to this day. In retirement, July 4th of 2014, I did not stop studying the matters of inspection and cleaning. Unfettered by a 'job', I could now develop in a non-commercial fashion! This not only meant there no longer were sales goals, but also I could imagine new ideas without other concerns. My search was for "best practice" and better methods and procedures.

With most all commercial efforts follows certain standards. Some are written and other implied by trial and error, word of mouth, or, practical apprenticeship. For me, the earliest training came from hundreds of technicians who would participate in annual sponsored "training seminars" for Bell South and Verizon. It was here that I was able to establish a life-long bond to "the craftsmen" who actually did the work. While someone in marketing always had an idea or two, it was the actual end user who influenced my thought processes then, and to this day.

What Was Wrong?

Two things troubled me the most about 'the tools of 1998' that are still in use today. The first was the way the actual surface is inspected. The second was the actual ways and means these incredibly small surfaces were cleaned. These interactive aspects are fundamental to anything that is cleaned: but were, and are largely ignored when it comes to the amazing sciences of fiber optic transmission.

Nearly four decades of training 'how to clean surfaces' pointed my search about fiber optic inspection and cleaning as being more 'true than false'. Still, there were gaps in how the fathers of fiber optics actually cleaned the surface. Their transmission science was amazing; their understanding of precision cleaning 'good enough but not so much'. This was because the means to "see the surfaces" was limited to a small area of the actual surface which international standards considered in 'zones'. These standards were written by the producers with little room for outside and independent input. Underlying this was a belief that "it's good enough for now": no one seemed to be looking to the future. It concerns me how many have been trained 'how to inspect and clean', over the last 30 years, with no inclination to "future proof".

Does It Really Matter?

Likely, there is a laboratory and scientists somewhere who are working on a thesis that a soiled fiber end face does not impact transmission rates! However, even if that is 'true', the reality is there are billions of heritage connectors where debris will impact insertion loss or misalignment will mean signal degradation. This study is a new base line that extends to all facets of fiber optic transmission. There are some in fiber optic transmission segments who believe that their end face is more mission critical than others! The reality is that be it fiber-to-the-home, or fiber-on-an-aircraft-carrier, trans-oceanic or a simple fiber loop in your community, or, a university campus or traffic control...the concept of "fiber hygiene" ... cleanliness of the surface, adequate methods and procedures, assures the future of this marvelous communications medium.

No matter the deployment type, there is one common thread and that is the condition of the fiber optic transmission "core" not only at the time of "test", but also in the time of "post test". Those are the topics of this work. Your input is encouraged as you have helped develop these networks and I am in debt to each of you who have taught me all along the way. We may not agree, but at least we think together.

The History of Telecommunications

It's not clear who invented the first telephone. Sometimes called 'the lovers phone', this device is a string suspended between two tin cans!

Actually, the history of the telephone is far more scientific, complete with immense science and intrigue. The "Father of the Telephone" is, of course, Alexander Graham Bell. He patented the device in 1896 after those now famous words to his colleague "Mr. Watson, come here, I want to see you." Most agree that the telephone evolved from Samuel Morse's 1835 invention, the telegraph, which transmitted signals by wire. Bell, at first, focused on improving the telegraph. As well, in 1874, Guglielmo Marconi developed transmission by radio waves. I find it fascinating that in these times these competing technologies of 'wireline' and 'wireless' still exist. In 1876, another inventor, Elisha Gray, worked to improve on Morse's invention with the idea, as with Bell, to actually speak over these wires. These is intrigue: in August, 2002, The United States Congress recognized Antonio Meucci as the inventor of the telephone. By that time, Bell's AT&T companies had expanded and advanced the telephone around the world.

The history of fiber optics also credits Alexander Graham Bell with an 1880 invention, the "photophone" which enabled voices to be transmitted by light. In the 1850's, British researcher John Tyndall proved that light could be bent in experiments passing light through streams of water. From that the 'game was on' as Austrian scientists Roth and Reuss used bent glass rods to illuminate body cavities for surgery. Between the 1920's and 1930's several inventors manipulated glass rods for rudimentary TV and facsimile machines. However, it was not until 1950's that Danish Scientist Holger Moeller actually contained a fiber core in cladding along with Americans von Heel and Hopkins. Hopkins and Narinder Kapany continued to evolve the science of transmission of fiber optic signals using laser as the light source. However it was not until the 1970's that Corning researchers Robert D. Maurer, Donald Keck, Peter C. Schultz, and Frank Zimar created the practical process for creating fiber optic cables we use today.

In 2017, Corning announced it had produced one billion kilometers of fiber optic cable!

As is the case with the telephone, there are multiple individuals who placed bricks in the wall that has lead to the current and future deployment of fiber optic networks. As you will read throughout this book, there is also a history of the sciences of Geometry, Physics and Microscopy that have influenced my study of the Art, Craft, and Science of fiber optic precision cleaning and inspection.

For me, this is a wonderfully delightful story of many who have brought us The Internet of Today and Tomorrow, Wireless Sciences, and, a future that is clearly unclear as to how creative and innovative is the human mind with sense of intellectual curiosity that has benefitted mankind and will for future-times. As is the history of the telephone and fiber optics relevant, so is the history of precision cleaning.

Because There Was No Proof:

In Spring-2016, as I was working on a White Paper, I realized that there was no proof of my thesis about the three-dimensional nature of connector surfaces or debris. Actually, an interferometer proved that debris was three-dimensional, however, the instrument only considers a two-dimensional surface...and the cost of even a modest interferometer is heart-stopping! There was no means to view connector SURFACES other than as 'characterized' by IEC 61300-3-35 and those standards that follow that baseline.

Video inspection utilizes a video camera and I began work on an instrument that employed a digital image camera...as on your cell phone! A patent was filed in Spring-2017 on means to connect the camera to not only a connector, but also adapters. The result are images you will see throughout this book. As of January, 2018 there are more than 3,000 still digital pictures and motion-video to 'characterize' all sectors as defined by existing standards. Additional 'zones' are considered to complete a three-dimensional 'real world' view that advances the science of precision inspection to a higher standard.

Enjoy the book! There is an MP4 version. Your input, critical to positive, is encouraged!

Prologue:

Caution: This book and contents are going to be controversial!

When I arrived at the fiber optic plant in West Lexington, South Carolina, it didn't look differently that the hundreds of other production lines I had seen since starting work in the Electronics Industry in 1971. My time at Union Carbide and Prestone® ended. My new career was in an emerging market sector: electronics.

As a 'manufacturer's rep', the firm sold and marketed two segments: one was Industrial Products that included test equipment and electronic components such as 'new fangled transistors'. The other side of the business included consumer electronics such as high-fidelity, portable radios and TV, and the wizardry of hand-held calculators. A major 'problem' in those times was reliability of staples such as TV and radio. Car stereo didn't work all the time. *Can you imagine?*

From this time I gained a foundation of sales and marketing that carried forward into my 2nd entry into electronics and fiber optics. I learned that it meant to be a 'manufacturer's rep': an invaluable assets to both start up and established businesses. I recall Mr. Sony speaking at an early CES (Consumer Electronics Show) as he defined the value of the manufacturer's rep. He said that 'the rep', as a means of sales and marketing, was viable to businesses whose turnover was about $200,000,000. Wisdom from this Master long before those who engineered Apple®, Amazon® and Facebook®...impacted my career then and now.

The Pirelli Electronics Facility in South Carolina had some of the brightest engineers I had ever met. They were on the leading edge of this new transmission science. At the time, there was exceptional quality control as finished printed circuit cards were installed. Fiber optic connector surfaces were 'bare open' and cleaning was a critical topic. Here the newly developed 'reel cleaners' were the standard means of cleaning, along with case after case of ubiquitous green boxes of paper tissue wipers. The problem was obvious: the cleaning materials and actual procedures were working against the need for a pristine surface. My report had the line: *"...they are doing it all wrong...."*. The problem was that by this time the fiber optic industry had established favorites and there was little room for new-comers!

One of the concerns was back-plane cleaning. Existing swab tools had adhesives and a balk ring that meant the swab could not be passed through the alignment sleeve. This meant the cleaning process could re-contaminate. The first patent had consideration for cleaning the alignment sleeve and passed through so that cleaning was a one-way track that would not re-contaminate. The second and third patents featured larger cleaning surfaces so that debris could be moved away from the initial point of contact. These products could be used "dry", but at that time a new phrase was coined and trained: "wet-to-dry cleaning" was included in various patent filings.

By 2005 Verizon began deployment of Fios® and it was apparent that existing cleaning methods and procedures were lacking. A better way to clean was developed that included use of non-IPA fiber optic precision cleaners. The actual technique became a generic term used to this time, but usually not as defined in the patents.

I have devoted these initial pages to "history" because the words of author and philosopher George Santayana are always important to me...and hopefully you! *"Those who cannot remember the past are condemned to repeat it."* This work is a history of science and practical applications for the future. It may be controversial because it speaks of a need to update existing standards for fiber optic cleaning and inspection that have commercial momentum of more than 30 years.

While some claim (incorrectly I believe) that their connector type is more critical than an other, the reality is that: aerospace or FTTx, long haul or metro loop, the fiber core is the 'common-thread' and most critical consideration...no matter the deployment.

There is significant history to the Science of Cleaning that began with the invention of soap in Babylon about 5,000 years ago! The sciences of magnification and physics are somewhat newer but these disciplines preceded the telephone and fiber optics by nearly 2,000 years. I hope you enjoy this book and the new science of precision cleaning and precision inspection.

Please join me: **In Search of: Best Practice**

"A ***best practice*** * is a method or technique ... that produces results ... superior to those achieved by other means...."

As a hobbyist race car designer, builder, and. driver for nearly 50 years, I grew to understand the term "Best Practice"! It may have taken a year or so of blown engines and broken axles to understand the stresses of wheel-to-wheel SCCA National road racing competition. The lessons were important to the hobby, as well as all endeavors in my life! This includes fiber optic precision cleaning and precision inspection.

Early in this 2nd Millennium I would be introduced to the folks at BellSouth and Verizon: "Best Practice" meant exceeding or even ignoring published standards. Both conducted rigorous annual training sessions and it was here that I first began to learn about deployments, real world trials and difficulties. These not only included 'day-to-day' matters, but also issues of training. At times the message was clouded by resistance to change, and in others, new procedures were readily and easily accepted.

Some consider the term "Best Practice" as a buzz word. Others dropped the title on anything and everything which tended to diminish the strength of the thought!

"Best Practice" has definition and it's found in major dictionaries and, of course, in Wikipedia®.

This work is "Best Practice" and it is based on experience and science. Best Practice is not something that comes and goes with time. Best Practice is a learned skill and trait. It's an ongoing thought process. It's the 20th of January, 2018: "Ricky", a uVerse® technician, just finished working on a problem. Five of his colleagues and countless conversations with technical services at Dell®, Microsoft® and Hewlett Packard® struggled. When "Ricky" walked in he had the same words as he departed: "I don't give up".

That is "Best Practice". <u>What is the Authority</u>? is asked throughout this book. For me, 'the authority' has been your experiences, which have challenged the new science of fiber optic transmission, science and history as underlayment of this ongoing study. You will learn about *The Scientific Method,* as defined as early as the 11th Century. I have used The Scientific Method to study and report these results. This work continues as our science not only transmits, but also, advances at "The Speed of Light".

"Making the Mazda Crazy" Road Atlanta

Fact or Fiction
of
Fiber Optic Cleaning and Inspection

As we begin this study time, let's think about some of the Facts and possible Fictions of a fiberoptic deployment. These are not presented in an order of magnitude and while all will be discussed in this book, there are others.

I suggest you write notes in the margins or create a notebook. Please, as your thoughts are not answered, or questions and challenges arise, I hope you will send me an email or give me a call. I have learned from you. You are where fiber optics "happen" and oftentimes your "raw information" is lost in a memo or conversation that doesn't make it up the line!

So: here they are.

1.) **Debris on an end face, and anywhere is a two-dimensional structure measured in diameter.** ___True ____False _____Unsure

2.) **Existing standards, such as IEC 61300-3-35, are "Best Practice"**
_____True _____False _____Unsure

3.) **Cleaning the fiber optic surface is important and anything is better than not at all.**
_____True _____False _____Unsure

4.) **A Fiber Optic surface is a two-dimensional structure**
_____True _____False _____Unsure

5.) **Cleaning for fusion splice prep is the same as end face cleaning**
_____True _____False _____Unsure

6.) **99.9% "Reagent Grade IPA" is an effective fiber optic cleaner**
_____True _____False _____Unsure

7.) **"Pass-Fail" automatic contamination detection is adequate to determine if connector surfaces are actually clean**
_____True _____False _____Unsure

8.) **There are many tools, including direct view and video inspection, that can determine if fiber optic surfaces are clean.**
_____True _____False _____Unsure

...with these as a background, let's study fiber optic precision cleaning and precision inspection.

What you will study may contradict current <u>trends</u> and what you've been taught.

**Let's separate Facts from Fiction
and
Myth from Scientific Reality**

What you have been taught over the last 30+ years may not be the same thing!

WHAT IS THE "AUTHORITY" ?

PRECISION CLEANING AND INSPECTION ISN'T "NEW".

- **'Wafer Fabs' for electronic components** (LEDs, CPU, RAM, etc....)

- **High-density printed circuit boards** (Pick & Place)

- **Metal Cleaning** (Aerospace, Nuclear, Gas & Oil)

- **Implantable medical devices** (Cochlear, Cardiac, Prosthetic)

A well-established performance history of precision cleaning and inspection on which our Industry can base itself...and has not.

Since other industries have advanced and evolved in their understanding of precision cleaning...why not fiber optics?

The reality is our deployments are as critical as work in a Class-1 clean room. Except that: we work in a dust storm at the top of a bucket truck in Kansas, or deployment in a military theatre, middle of the ocean or Amazon river Basin!

What Does It Mean? "Future-Proof"*

There is another term we hear in seminars and read in White Papers!

"Future Proof" has a formal definition and it begins, for me, with the quote about 'history' and lessons not learned'!

1. **Processes that anticipate future developments.**

2. **Actions taken to minimize negative consequences.**

3. **Change old perceptions into new realities**

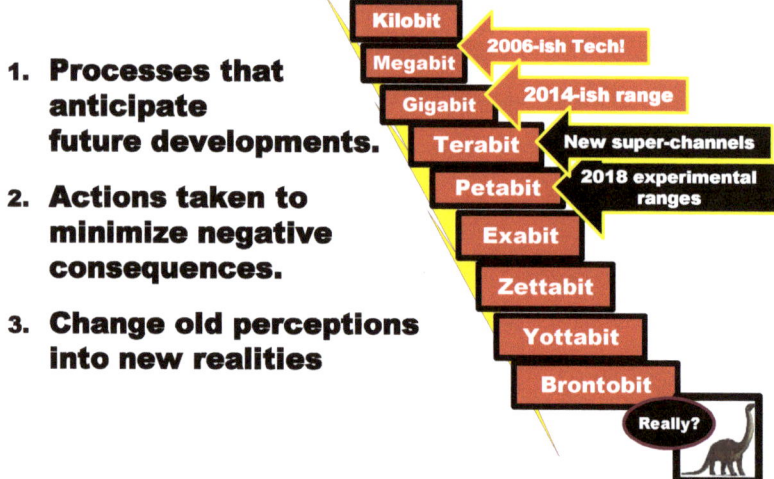

THEORETICAL TO PRACTICAL: Our Industry moves practically at 'the speed of light'! It was only ten or so years ago that banners at trade shows touted "megabits" over "kilobits". Now our scientists are studying Petabit transmissions and means to increase capacities.

Some of you may see a "Brontobit" in your lifetime but the topic of "Future Proof" not only relates to these fanciful speeds and capacities. The most important Future Proof is your training.

It's essential that we pay it forward by continually understanding not only the sciences of transmission, but also, how precision cleaning and precision inspection impacts the future of the Industry. This means we all anticipate developments to minimize consequences and this happens by changing old perceptions into new realities.

* adjective: British. Oxford® Dictionary. Technipedia®, Cambridge® Dictionary, Wikipedia®

Buzz Words?

Future Proof

Best Practice

Or: **The Way Forward**
"Do it right the first time"

"Do all aspects of Fiber Optic Network Design, Installation and Training Development?"

advance at

The Speed of Light?

"Maybe..." *Is that good enough?*
The "maybe" is also a topic of this book and MP4 seminar.

Myth or Reality?

What is the "end face"?

"Zone-1" ... is the "fiber/core"
"Zone-2" ... is the reflective cladding
"Zone-3" ... is the contact surface

Is the IEC 61300-3-35 (rev-2) definition of a two-dimensional surface adequate and accurate?

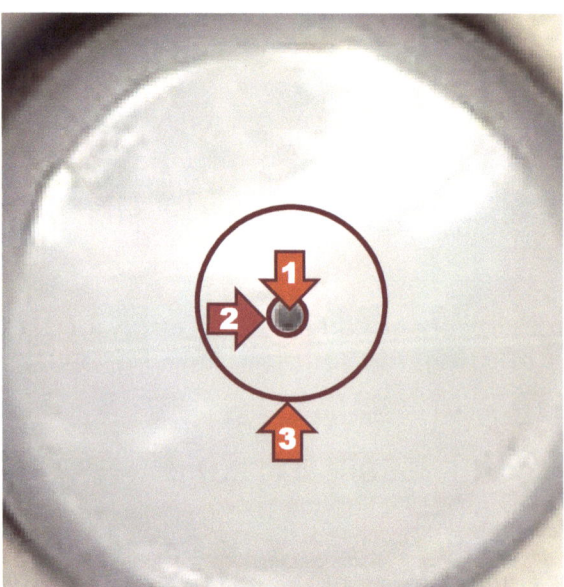

THE "MYTH" OF TWO-DIMENSIONS

This portrait is a digital photograph of a 2.5mm connector surface. On the next page is an image of the same "end face" that conveys the 'virtual 3D' nature of all fiber optic surfaces.

Some areas you see and others you do not.

Myth or Reality?

The reality of 3-D

It's unlikely you have seen an image of a 2.5mm connector that portrays not only the 'standard end face', but also other aspects of the connector surface.

In April, 2016 I began work on a new means to inspect these surfaces. This is digital photography and the instrument (patent pending) has captured more than 3,000 photographs of all types of fiber optic surfaces...and contamination in 'strange places' other than the 'defined standard view'.

The reality is that debris can be present on all sectors of the connector as well as adapters and alignment sleeves. This is true for all fiber optic connectors in all applications.

Myth or Reality? The fiber optic end face and associated components are 3-D structures

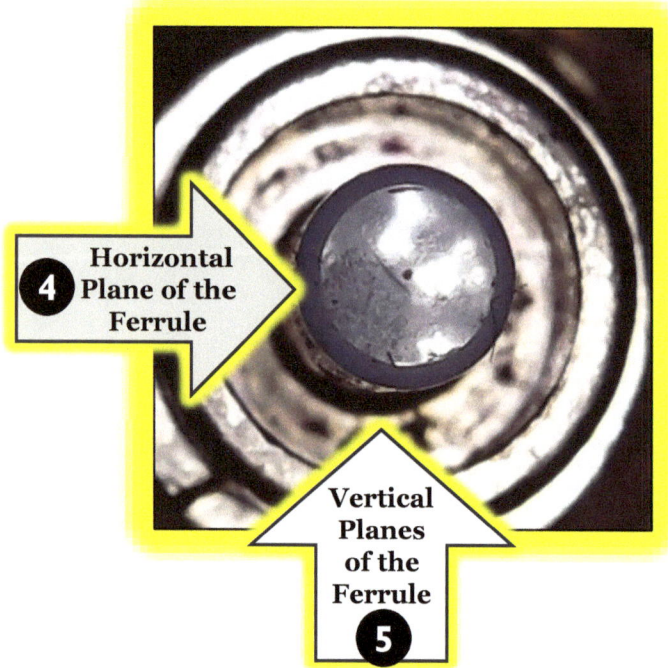

4 Horizontal Plane of the Ferrule

5 Vertical Planes of the Ferrule

Add Zone-4 and Zone-5 to your understanding of any fiber optic surface

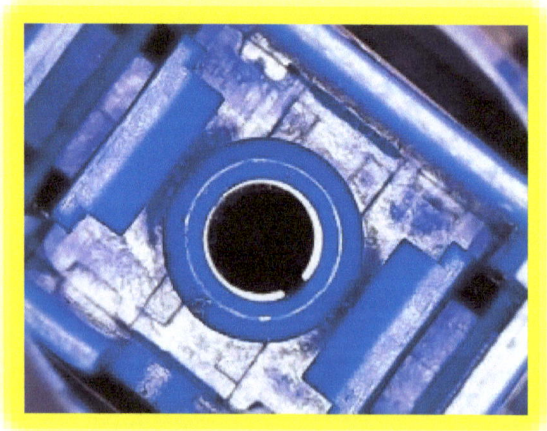

Images by RMS-1©

Include these surfaces

How does contamination "get there*"?

These things "happen"! Bottom left is a jumper with "condensation contamination". It was about 22F in Pittsburgh and I was scheduled to lecture in about an hour. I moved my equipment from car storage into the meeting room. As I set up the picture (1) appeared on my ME-9600 standard video scope. Frankly, I was 'horrified'...never had I observed condensation contamination.

Some years later as I was testing RMS-1™, I captured the image in (2). The "core" and "horizontal surface" was clean...but this piece of thread was on the "vertical surface". I thought to myself: *"There is no way I could have staged this 'proof' of debris on the area not seen by standard inspection."*

The image to the right (3) is one that many of you have demonstrated to me! This is "transfer contamination" from the jumper side to backplane.

"Contamination Events" happen! Since most inspection can't see these surfaces, or we are not prepared to be aware of these types of debris...we just don't know! Standards or test gear doesn't characterize these surfaces.

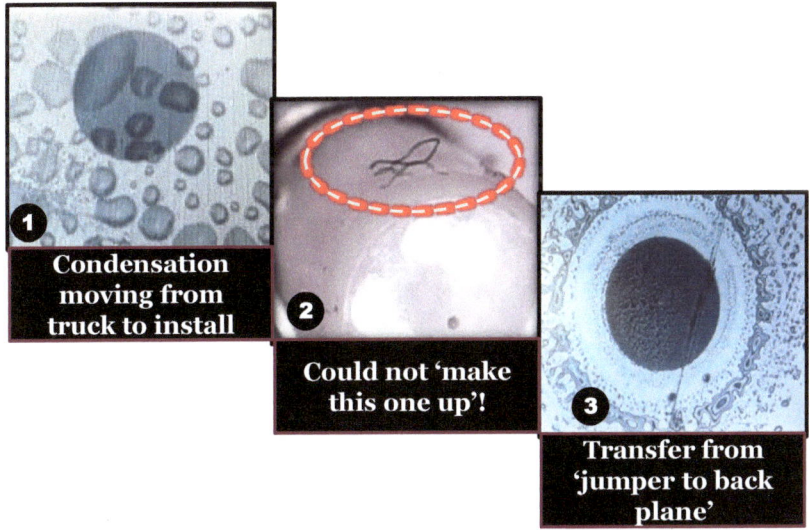

1. Condensation moving from truck to install
2. Could not 'make this one up'!
3. Transfer from 'jumper to back plane'

Sometimes, it's "impossible" grasp the vast and infinite nature of our universe of fiber optic installations!

We want "it" to be 'easy' and 'convenient'. What is "it"?

"It" is the design, installation and training process. Oftentimes, these functions are not interactive. For example, a manufacturer will state in installation instructions to 'clean the surfaces' without clear applications specific directions 'how to clean'! The result can be a warranty claim…or another jumper installed and one that could have been cleaned trashed!

There is an assumption that 'everyone' knows how to clean! The reality is, as you will read on, the instructions to clean can be vague and counter-productive.

My recommendation is that a network designer should include specific cleaning and inspection instructions with each document. Who better understands the actual physical environment, complexities such as time of year when there may be a monsoon or pollen, or, the skill level of the actual work force?

Trainers have a special role. While your specific skill may be honed for one piece of equipment or installation group, there is also a need to jump-back-in to that PowerPoint® and update your slides or script.

Each of us bears a special responsibility to assure the medium of fiber optics is successful. In my career I experienced "sure bets" such as VHS® and BetaMax®, 8-Track and Cassettes, 10-key machines give way to scientific calculators, fax machines, vinyl records …
all replaced by unforeseen technology.

It's not likely, but possible that something as 'simple' as a contaminated fiber optic connection can make this important advance as relevant as a waterbed!

You will learn from this session that cleaning and inspection can be 'convenient-and- easy'…*"it" can also be accurate and effective.*

Who are these people and what could they possibly have to do with fiber optic cleaning and inspection?

...A brief lesson in the History of Geometry and Physics...

We began this study with an understanding of the history of the telephone and fiber optics. So others might understand 'conclusions' are not derived from a 'sales and marketing story', let's understand a little about the sciences and how humanity arrived at something we take for granted: existence is three-dimensional!

Euclid of Alexandria is considered "The Father of Geometry". About 2,500 years ago he defined shapes in two dimensions: length and width. It's not easy for us, in these times, to consider there was a time when 'science' began to unwind from mystic-crystal-revelations to hard proof! His thirteen text books, Euclid's "Elements" were a rigorous proof of mathematics. It is said that he did not simply want to 'feel good' about his study, but to demonstrate it as proof. <u>*"The laws of nature are but the mathematical thoughts of God"*</u>: Euclid.

I continue to conduct live demonstrations of cleaning procedures some of you may have witnessed. For others, these are also available on YouTube®! (YouTube Channel ID: UC1a552-2i62oUP6mM9WhwRg)

In 80BC[ish] the Greco-Roman astronomer-geographer-philosopher Ptolemy looked at the heavens and realized there was "distance". He created a universe that considers the planets and sun as revolving around Earth. This belief was accepted until the mid 14th Century when Nicolaus Copernicus re-wrote the understanding, ultimately resulting in Christopher Columbus sailings and those who ultimately circumnavigated the planet proving it as a three-dimensional sphere. There was a time when we were not as smart as we are now!

In the 11th Century, while Europe was in The Dark Ages, Islamic scholar Ibn al-Haitham studied eyesight and lenses. His work laid the foundation for 'The Scientific Method" and was widely recognized during The Renaissance. The Scientific Method is used to this day: it's a foundation for this study.

In the 17th Century, Descartes expanded on Euclidian geometry with his work on spheres ... 3-D science we know today. Rene Descartes is considered the Father of Western Philosophy and as such, continued Euclid's thesis of "demonstration" to remove doubt.

These foundational scholars, some with thesis that are accurate in today's world and others 'not-so-much' all formed the basis for reality based science. Euclid, Ptolemy, al-Haitham, Descartes, Copernicus and even Columbus were considered as 'polymath' (a sage) or mathematicians until the early 20th Century when two other scientist-philosophers added to the sum total of humanities understanding of existence!

Euclid, Ptolemy, al-Haitham, Copernicus, Columbus and Descartes formed the basis for the work of Physicists Albert Einstein and Max Planck.

Early 1900's

Planck's *Quantum Theory of Physics* revolutionized our understanding of Atomic and Subatomic structures; Einstein's *Theory of Relativity* revolutionized our understanding of space and time.

Max Planck's work on sub-atomic structures defined them as 3-D ... and tied Euclidean mathematical science to physics! His work broke down particles to the smallest levels and defined these as three-dimensional structures. Planck also studied light including work with electric companies to better the light bulb! Planck received The Nobel Prize in 1918 for his study of sub-atomic structures. As with those before him, the quest for proof extended beyond the concept of "feels good"!

"Stay with me here for a few minutes...this is 100.5% relative to fiber optic inspection and cleaning!"

Albert Einstein received The Nobel Prize in 1922, Albert Einstein was awarded the 1921 Nobel Prize in Physics, "for his services to Theoretical Physics, and especially for his discovery of the law of the photoelectric effect". This tied together many, if not all, of the sciences of historical sciences of transmission. An Einstein quote: "You see, wire telegraph is a kind of a very, very long cat. You pull his tail in New York and his head is meowing in Los Angeles. Do you understand this? And radio operates exactly the same way: you send signals here, they receive them there. The only difference is that there is no cat." In his later years Einstein turned his thoughts to philosophy.

You are urged to "Wiki" both of these individuals to understand how their contributions influence our work and thoughts in these and future times.

These scientific advances, only barely defined here, form the basis for inquiry I used to study precision cleaning and inspection. The rigorous efforts of these monumental minds are a base line for all who study and advance to "future proof" by installation of "best practices".

...about the year 1590, two Dutch spectacle makers, Zaccharias Janssen and his father Hans started experimenting ... *put several lenses in a tube and made an significant discovery.*

THEY HAD JUST INVENTED THE COMPOUND MICROSCOPE...
PRECURSOR TO CONTEMPORARY VIDEO INSPECTION

As we have seen before, the history of 'who' can be clouded. Some historians say it was Hans Lippershey, most famous for filing the first patent for a telescope. Other evidence points to Hans and Zacharias Janssen, a father-son team of spectacle makers living in the same town as Lippershey. The Janssen's work employed multiple lenses to create a 'compound microscope'.

Galileo based his devices on their work and added a focusing device and continued to explore the heavens with his telescopes. He described the principles of lenses and light rays and improved both the microscope and telescope.

Dutchman Anthony Leeuwenhoek worked with magnifying glasses in a dry goods store and used a magnifying glass to count threads in woven cloth. He became so interested that he learned how to make lenses. By grinding and polishing, he was able to make small lenses with great curvatures. These rounder lenses produced greater magnification, and his microscopes were able to magnify up to 270X!

Anthony Leeuwenhoek became more involved in science and with his new improved microscope was able to see things that no man had ever seen before. He saw bacteria, yeast, blood cells and many tiny animals swimming about in a drop of water. From his great contributions, many discoveries and research papers, Anthony Leeuwenhoek (1632-1723) has since been called the "Father of Microscopy".

The fascinating history is not unlike the questions of the actual inventor of the telephone or recent attention to Nicola Tesla as inventor of the radio, X-rays, and, fluorescent tubes.

My point in the detail of this section is to create a historical time line so we can establish the foundation all of these brilliant contributors used to prove the thesis for fiber optic applications in the same scientific sense as proven and employed for more than 2,000 years.

THESE HISTORICAL AND PRACTICAL REALITIES ARE ESSENTIAL TO PRECISION INSPECTION AND PRECISION CLEANING FIBER OPTICS!

1. Existing precision cleaning and inspection standards are based on "Euclidian 2-D geometry".

2. Ptolemy, Descartes, Planck, Einstein and (even) Christopher Columbus taught us ... *we know as you are sitting there today in the here & now:*

3. **WE EXIST IN 3-DIMENSIONS!**

IT'S TRUE:

We are <u>inspecting and cleaning</u> fiber optic connections based on 2nd Century (BC) Euclidean Geometry and Microscopes invented in the 16th Century!

It's time to ... "**GO 21st CENTURY**" ... with our thinking about these things!!!

Mini-Conclusion:

Myth or Reality?

THE "MYTH OF TWO-DIMENSIONS"
Existing standards define a fiber optic end face in diameter as a two-dimensional horizontal structure.

Connectors are NOT two-dimensional structures as defined by existing standards in "diameter" and characterized for "pass-fail" analysis.

You don't need these folks, or, a Physics and Geometry lesson to explain: "common sense"!

"Wait...THERE IS MORE"
Existing Standards Measure Debris & Contamination in "Diameter of 2-D"

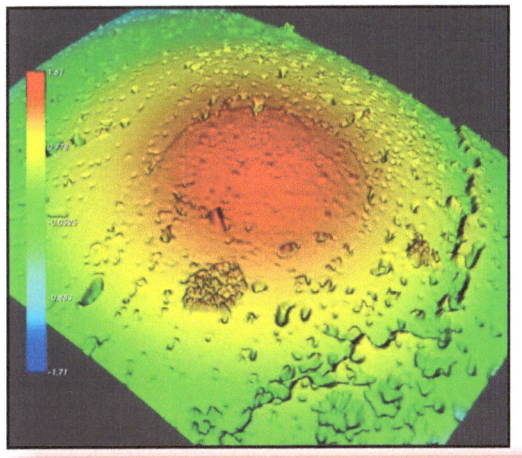

The "red" on this interferometer scale is (max) 1.67 microns in *width and height.*

SET THE RECORD STRAIGHT!
Contamination, in any or all forms, is NOT two-dimensions.

You are NOT sitting there in two-dimensions!

Connectors and Contamination are NOT two dimensions as well!

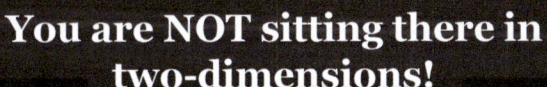

A "Homer Simpson® Moment?"

Who has the 'need to know'?
AN INDUSTRY-WIDE UPDATE IS IN PROCESS

Mini-Conclusions:

"WE" ARE STILL EVOLVING.

- Copper continues to grow even in the face of replacement by fiber optics.

- Fiber Optics is challenged by wire wireless, DOCSIS 3.1. Who knows what 5G will mean to all of us!

- You as a network designer, technician or trainer are challenged by: newer, faster, better.

As have other sciences, we also have had false-starts. Might be inspection and cleaning?

No matter the application or connection type ... when deploying fiber optics ... there is an end face and associated connector geometry that must be cleaned and inspected ... properly.

Future generations will benefit from what you do now
Design-Install-Train-Manufacturing

We all share in the responsibility to advance
The Sciences of Fiber Optics to an ever-higher level.

The Myths of Test and Inspection:

Before you clean, how do you know what debris you are cleaning?

After you clean...how do you know...the surface actually clean?

For some, there is confusion about which test instrument actually can determine a clean connection.

SET THE RECORD STRAIGHT: WHAT DO I USE?

- **Fiber Identifier**
  - **Visual Fault Locator**
- **Db loss test set** (power meter & light source)
  - **OTDR?**
- **Direct view microscope or loupe? A magnifying glass?**
  - **Video Scope?**

The "real question":
What does it do?

Is an important topic because
"I tested it" ... should be clarified "...with what did I test"?

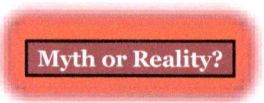

What does it do?

Fiber Identifier: **A Live Fiber Trace and Toner** _enables the technician to identify available fibers_ without the disruption of existing subscriber services.

The Fiber Identifier will not indicate a clean/soiled surface

A live fiber may not be a clean fiber!

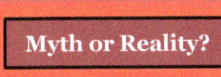

Myth or Reality?

What does it do?

Visual Fault Locator: **A visual fault locator *uses a high power visible laser designed to locate and <u>identify faults and breaks</u> in fiber optic cables, patch panels and other cable splice areas.* They are typically effective up to 7km.**

- **This instrument <u>will not</u> determine a clean/soiled fiber**

What does it do?

Db loss test set (power meter & light source):

Provides insertion loss measurements...*how much light is lost on a fiber 'run'*

○ Will <u>not</u> determine a clean/soiled fiber

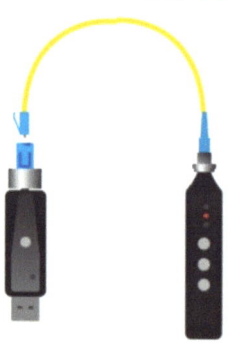

Gives accurate readings of loss, *but won't tell you "why" or "how"*.

The Myths of Test and Inspection:

What does it do?

OTDR? Provides a graph of a fiber optic cable typically reporting:

- Distance to faults and connectors/loss per km
- Length of the fiber run
- Verify splice loss
- Reflectance and Insertion Loss

The OTDR is a "beautiful thing"

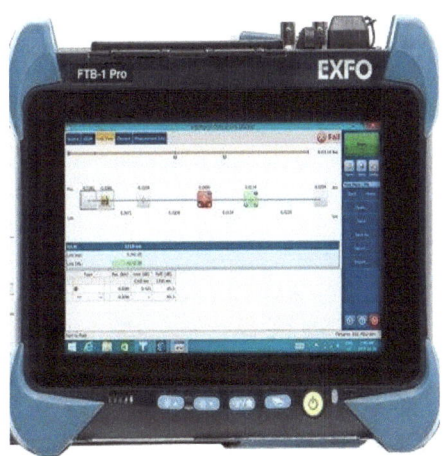

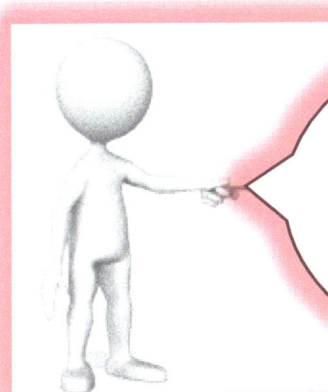

"Measurements are possible, using an OTDR, that can indicate a clean or soiled connector surface.

It's a lot of gear just to indicate clean!"

Myth or Reality?

A FIBER OPTIC VIDEO SCOPE is

The "Best Practice", most-practical and scientific means to assure fiber optic surfaces are precision clean…

Which one works best?

There are many types of fiber optic inspection. Most are video cameras of varying magnification that are aimed at the surface of the fiver connector as defined by existing standards. There are new 'convenience' instruments that record and analyze in 'pass-fail' modes.

These devices are based on microscopes and telescopes patented in the 15th and 16th Century. Some of these are actually direct view microscopes. There is a new concept that utilizes digital photography.

I characterize these instruments into four classes.

It seems we always ask: "Which is best?" What if each had a place in test, measurement and training functions.

Consider : "The 4 Classes of Inspection"

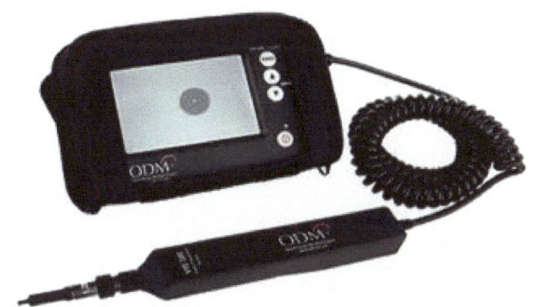

Class-1: Probe and Monitor

Readily available. Reasonably priced. Check for 'trade-in' if you are on a budget. Available as "bench top" and "portable".

Some are 400x and others 200x and lower.

The lower the magnification the more of the end face you will see. At 400x the viewed surface area is small.

Why might that be a concern?

Consider : "The 4 Classes of Inspection"

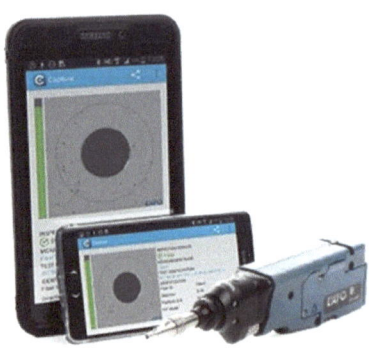

**Class-2:
Wireless/AutoDetect**

Amazing technology and convenience. Most are 400x and marketed to meet existing standards. Available as 'bench top' and 'portable'.

Priced on 'upper end'.

Analyses 'standard debris' by algorithm analysis of a limited end face area and limited debris type.

TechTip: Check to see if 'auto-detect' can be disabled

Consider : "The 4 Classes of Inspection"

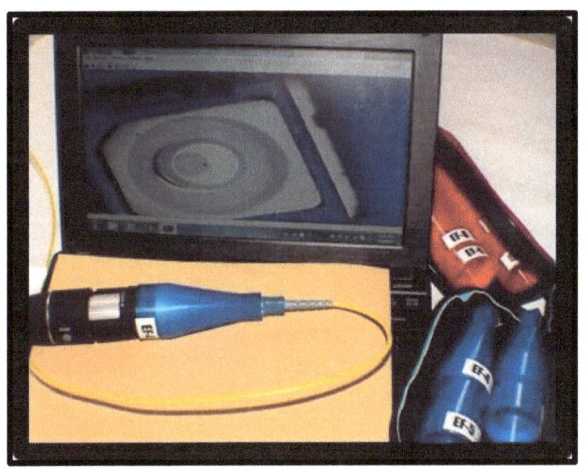

Class-3: Views 3-D Nature of any connector & debris

New concept features digital photography that observes all sectors of the connector. Exceeds existing standards. Currently only available as 'benchtop' model.

Up to 200x magnification. Captures wide range of connector surfaces in still and motion video, Images can be enlarged on PC (included)

Ideal for training and research

Consider: "The 4 Classes of Inspection"

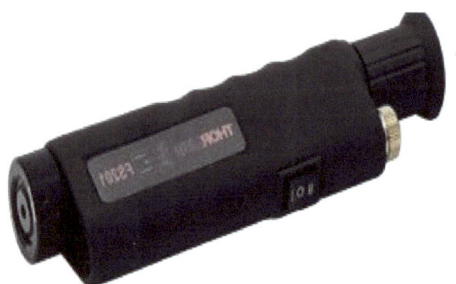

Class-4:

DIRECT VIEW MICROSCOPE OR LOUPE

- Descended from the 16th Century, the "direct view" microscope, *should never be used on an active fiber.*
  - "Far too often": Direct View microscopes are used *improperly and unsafely where* video inspection is compulsory
  - My personal feeling: *"Never Trust a Claim these are filtered".*
- Ideal: QC jumpers...

Which works best?

"It depends…" on your applications specific needs.

Don't be "sold" … listen to all the 'technology reasons' a specific manufacturer has designed into their instrument.

Since inspection is the most critical tool in your box, make sure you are making the right decisions. For example, you may require a scope for each technician and some models can 'break the bank'! Can a less expensive model be "best practice"? I believe that it's more important that each technician have a 'scope' than sharing one! Invariably, the unit is in Chicago when it's needed in Joliet!

The new models of "pass-fail/auto-detect" are amazing science. However, there is also craft and art that remains as an integral part of our work. The point is that you may be able to purchase two lesser quality and this be 'good enough'. All of these questions are important 'due diligence'. The most important consideration is inspection itself and integrating this into the work regimen.

Existing standards 'call out' 100% inspection of all connectors. When faced with a bank of connections at a data center or working in a rain storm this can be a challenge to even the best technician and well-run organization. For this reason, I have always considered a Plan-B.

"Plan-B is a cleaning procedure that works close to 100% of the time to parallel the goal of 100% inspection.

Myth or Reality?

Mini Conclusion

The "myth" of two-dimensions is embedded in fiber optic cleaning and inspection training and standards.

"Does it really matter?"

As we continue, and as we have learned from history, there is "science" to everything! I believe that existing fiber optic transmission sciences are beginning to out-pace our understanding of the sciences of inspection and the sciences of our understanding of contamination. Simply put, we have not had time to consider the many aspects of inspection and cleaning that have other industries.

The remainder of this book is written, in the same ways as Euclid and others studied their science, not to "feel-good" but rather to present the reality of precision cleaning and precision inspection.

Enjoy!

AN IMPROPERLY CLEANED CONNECTION CONTRIBUTES TO: Insertion loss as a result of reflectance or refraction of the signal. We speak about cleaning a connection, but also, an improperly cleaned connector also can result in a flaw test result. Before you conduct ANY fiber optic test, be certain you clean the fiber connector surfaces. Time is money…an improperly cleaned connector can also be a part of a lo$$ budget! Consider, with care, the ways and means you clean and inspect.

FIBER LOSS: Early scientists quickly learned that the light source was a "problem". Narinder Kapany, along with Drs. Mensah, Schultz, Keck and Maurer created cables and lasers that enabled greater transmission length and capacity. Light loss along the fiber is integral to the insertion loss budget.

SPLICE LOSS: is often considered to be a quality of the splice. However, the quality of the actual splice is not only a characteristic of an individual machine, but also its condition and preparation of the glass for the splice event. Always assure the operation functions of the splice machine are clean: guides and trays. Assure the glass is clean. You will read on how IPA degrades and use of this chemical can result in premature degradation of fusion splice electrodes.

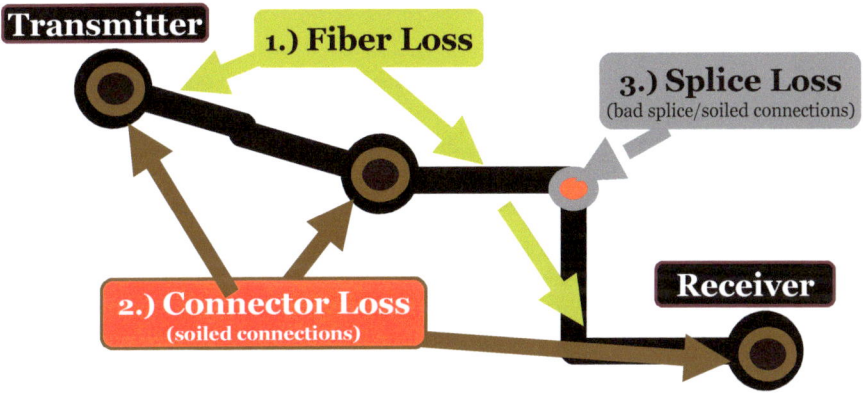

Insertion Loss is a 'design sum' total of three components.

It's an essential calculation in every fiber optic installation.

> There are many "opportunities" for insertion loss "transmitter-to-receiver".
>
> *In this session we'll review some you know... and...introduce you to others you may not!*

No matter where you work in our Industry, there are many opportunities! As well are the 'opportunities' that challenge your work! When you consider how many connectors or splices between transmitter and receiver, each and every $1/10^{th}$ of a db counts!

WHAT ARE "SOIL POINTS"?

Since each connector as a three-dimensional 'geometry', there can be many locations, other than the flat surface of the end face, where debris can reside. I term these 'soil points'.

Can you imagine how many connector types there are? Some of specific to military and aviation, others to data centers, and still others to campus environments and that broad category: FTTx! Some workers and SME will claim that their 'military connector' is better than a "mere" connector for fiber-to-the-home! Others realize that there is a common thread: the condition of the actual transmission fiber at the time of transmission.

Don't assume that since you understand one connector type that others are the same! For example, the 'soil points' possible on a "LC" are different than those on a "shuttered LC". Look at each connector type for commonalities and differences: explore best practices all the time. Share your knowledge on 'professional social media'.

In the following sections we will discuss fiber end face condition at the time of transmission. You will learn this may not always be the same thing as 'condition at the time of test'.

Understanding all fiber optic surfaces is critical to successful deployment

<u>Reflectance:</u> "Reflected Power" can be caused by a soiled or improperly cleaned connector. Acceptable loss on single-mode is ... maybe .2-.5db. Much is written about "reflectance" and this *is* a critical topic.

However, there are other causes of signal degradation and these can be caused by debris on surfaces other than characterized as the 'standard end face'.

Another phenomenon causing signal loss is 'misalignment'.

The red/yellow arrow points to debris trapped between the two end face surfaces.

The blue/yellow arrow points to the light path that has been skewed by this misalignment.

Misalignment is also possible if debris is imbedded in an alignment pin hole. There is dust on other 'soil-point' sectors of this "MT-Type".

The gouge through the fiber is obvious and a concern.

Of equal concern is debris at the base of the ferrule that can also create misalignments or be 'picked up' by the (repeated) physical insertion process of the connector.

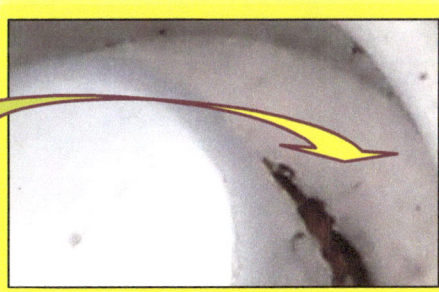

Mini-Conclusion:

Cleaning (all) fiber optic surfaces is critical to successful deployment

As Speeds and Capacities increase ...
...as well, so should your <u>awareness</u>
Three Dimensional Nature of:
a.) The connector ...
b.) ...contamination type
c.) ...and location.

Images by RMS-1©

Myth or Reality?

What is "Primary" and "Secondary" Contamination©

Not only are there infinite debris types, but also the locations should be characterized.

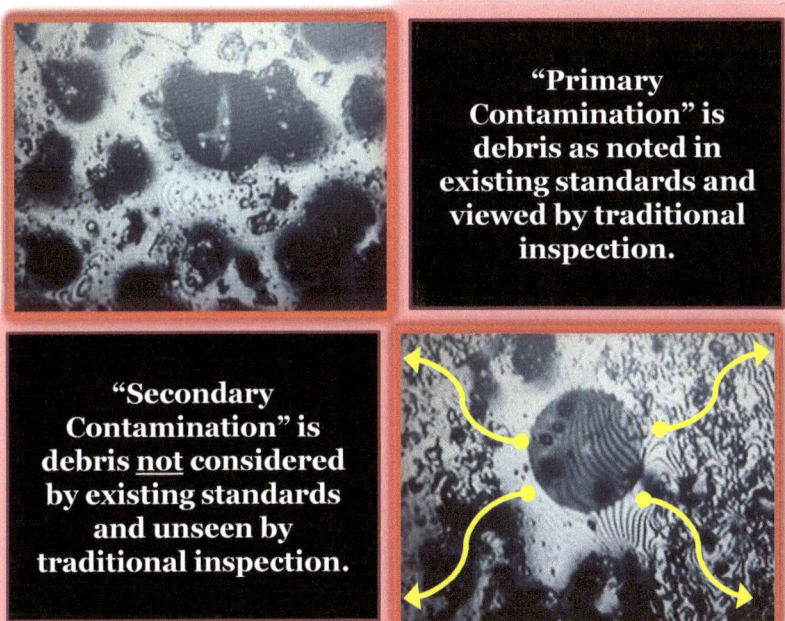

"Primary Contamination" is debris as noted in existing standards and viewed by traditional inspection.

"Secondary Contamination" is debris <u>not</u> considered by existing standards and unseen by traditional inspection.

What's Happening Here?

These are images of a 'fingerprint'. Upper right is the end face before mating, lower left is the same surface after mating. The yellow arrows depict how the finger oil has been 'pushed out' by the mating at the 'contact zone'.

However, this debris field does not end at the picture! The fluids transfer to areas not seen by existing microscopy and not considered by existing standards. Understanding Primary and Secondary Contamination is 'simple science' and common sense that results in best practice.

Myth or Reality?

Examples of "Secondary" Contamination©

"Secondary Contamination" may be present on connector sectors outside the 'field-of-view' of an end face as defined by existing standards.

<u>Image top:</u> debris on the surface of this alignment sleeve can be 'pushed' from the 'jumper side' to the 'backplane' side.

<u>Image center:</u> there is debris on the "horizontal ferrule" and embedded within the connector.

<u>Image bottom:</u> debris present throughout the connector structure.

What is "Primary" and "Secondary" Contamination©

"Primary contamination" is soil deposited on the end face surface by touch, environment, or other activity. *It is standardized by several organizations.*

"Secondary (cross) contamination" is debris of any type *outside the limits* of <u>most</u> existing inspection and <u>not often</u> characterized by existing standards.

Awareness of "Primary and Secondary (cross) contamination" provides a more complete understanding of connector, debris, and best practices for fiber optic design, installation and training.

It is a higher standard.

"PRIMARY" AND "SECONDARY" CONTAMINATION:
EQUAL OPPORTUNITY PROBLEMS!
(To a successful deployment)

➢ We all have been taught to clean (and inspect) The End Face: *that's a good start.*

➢ Now for the first time, we see proof there is more to consider than "only the end face".

➢ *This means that if you have cleaned multiple times with no positive result ...* you may be cleaning the wrong sector of the connector.

"What is the '3-D' nature of a connector?
Isn't '2-D' good enough?"

This is the 'standard IEC 61300-3-35 view' of an end face

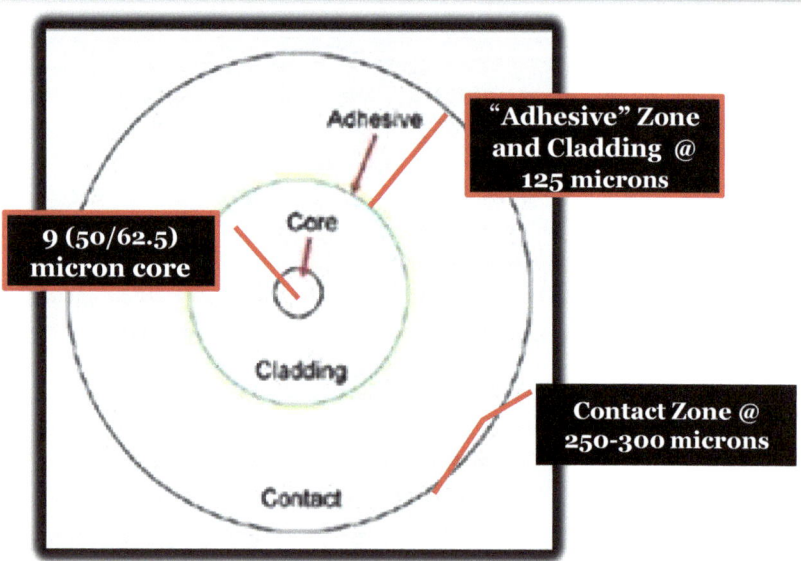

9 (50/62.5) micron core

"Adhesive" Zone and Cladding @ 125 microns

Contact Zone @ 250-300 microns

For most of you, this is a 'review' of what you have seen or been taught many times. There are "zones" often attached to these terms: "Zone-A" is the fiber "core", "Zone-B" is the cladding and sometimes the 'adhesive area' abutting the cladding is termed "Zone-C".
This makes the "Zone-D" the contact zone.

The real significance of these drawings is that the inspection and cleaning areas end at a point that is a small fraction of the actual surface.
What does that mean?

The other consideration I add is that this drawing is the same for both single mode and multi-mode fibers. Some like to separate single mode from multi-mode. I believe that since multi-mode speeds and capacities are being pushed toward single mode levels, the commonality of understanding is important. There are new converters to increase multi-mode to single-mode.

The reality is that our Industry is being serviced by technicians of varying skill levels. To me, it's easier and "Best Practice" to treat both single mode and multi-mode with the same attention to detail.

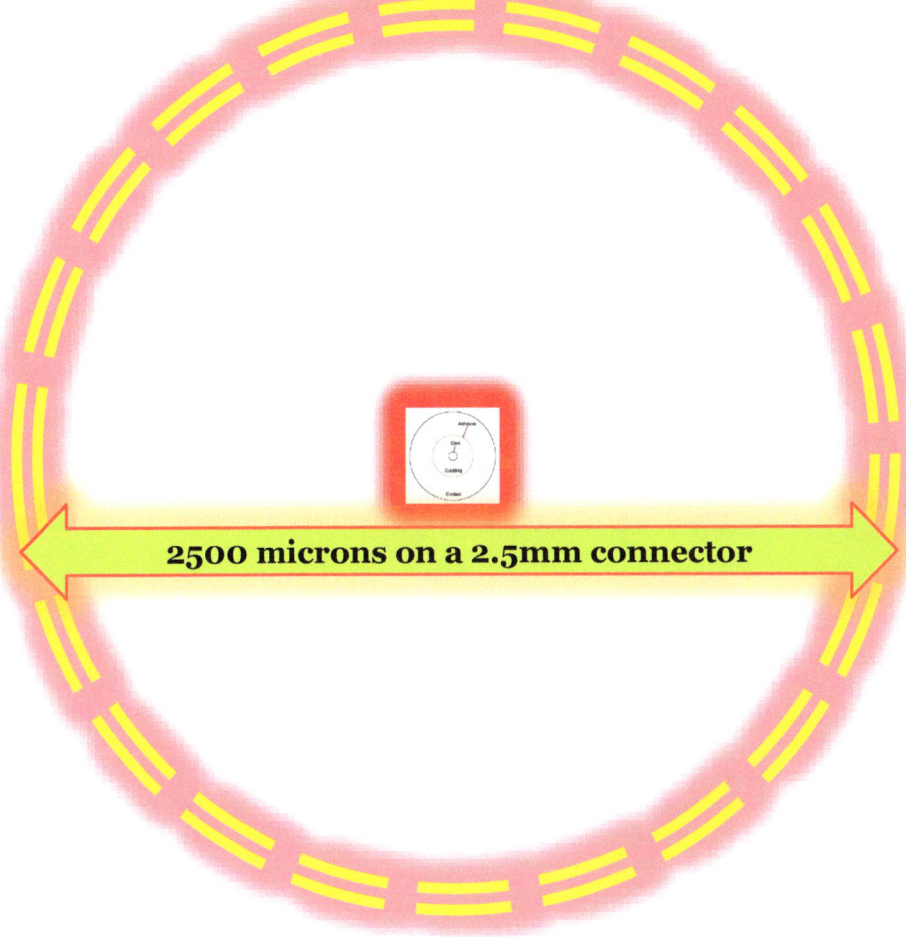

When you 'do the math' it's very revealing.

A 2.5mm end face is 2500 microns. The yellow circle is about 5" and the graphic from the previous page is ½".

This is a representation of how much of the end face is seen using existing 400x inspection. "What Happens", outside sectors that can be seen or characterized, matters!

This is a short video that demonstrates the complete "field-of-view" of a "FC" connector. It was taken with the RMS-1 scope and I will email you a copy on request.

As you look to the traditional end face, which actually magnifies 6x larger than this picture, there is limited debris on the "standard end face". The reminder of the connector surfaces is heavily contaminated ... which can cause "Secondary (Cross) Contamination"

Every fiber optic connector type should be characterized in this way.

**This video runs live in the MP4 version of this session. If you would like a copy, please send an email:
edforrest@fiberopticprecisioncleaning.com**

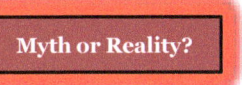

Myth or Reality?

This book, and associated MP4 presentation, is based on established science that may seem (a little) foreign to you.

Over the last 30+ years we have been 'taught & marketed' many stories. Today you are receiving new information that updates the last 30+ years.

As we move from "theoretical" to "practical", this <u>science</u> and <u>continual</u> update is important to the future times of fiber optic deployments of fiber optic types. No matter your specialty: FTTx, Military, Traffic Control or Security...long or short haul. All are linked by the condition of the fiber core at time of test and transmission.

Let's continue with a discussion of contamination. What is "debris", actually? Why is it easy to remove on some surfaces and not so much on others? When we speak about "residues", what is that, exactly?

Debris, contamination, and 'dirt-dust' are composed of two general types, and (literally) infinite subsets. There are ionic and non-ionic solids...these are sometimes called "polar and non-polar". The 'brief-story' is that ionic solids are electrostatically held together and non-ionic solids are chemically bonded. What this means to you is that some cleaning methods for one type, may not work well on the other.

For fiber optic precision cleaning, I considered not only various products to clean, but also techniques and procedures. The goal was not to 'feel good', but rather come to proof that one procedure would enhance probability of removal of all types...without the demand of trying to follow a flow chart or applications matrix!

At the end of the day, all we want is a job done right...the first time.

The first cleaning tools set the expectations of convenience and simplicity. In those times, meetings and events touted 40 megs per second. It was almost as though even the scientists of the time didn't imagine our capacities, and the demands, would pass from 'theoretical-to-practical' as quickly. At the time, the concept of FTTh evolved and the suffix "x" replaced the last letter as applications multiplied: traffic control, security, apartments, wireless cellular back-haul, and broadcast were added to long-haul trans-national and trans-oceanic routes. Add commercial aerospace, military and outer space applications to a telecommunications foundation that was, and continues to replace copper infrastructure.

To perform this work meant that contractors re-trained and evolved. Some eagerly accepted fiber optics and others stood pat and grew with the amazing advances with "Category Cable". Trade groups such as BICSI® and ETA®, as well as professional trade show managers accepted the challenges of re-training. What once was a 'carnival of new products' at shows such as OFC®, NFOC©, SCTE®...among many others, became formal training events to the point that vendors complained there were no attendees 'walking the aisles' as professional crafts persons re-educated to the disciplines of ultra-capacity!

New companies entered the market, trying to understand whether fiber optics was an 8-Track Tape or long-term institution that would, in reality, replace the lines that Alexander Graham Bell string with buck boards pulled by mule-power! While so many aspects of fiber optic deployment were clear...what still confounded the installer...was the question: "Why am I having such a problem cleaning these things?"

Myth or Reality?

Let's talk some trash!

Dry Debris tends to stay in place

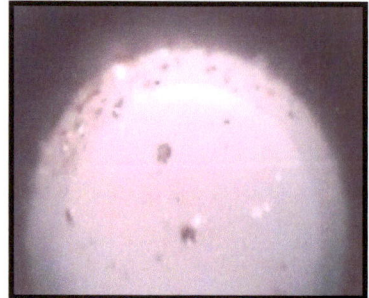

Fluids move and can transfer

Some contamination is unknown

There are three general categories of debris:

1.) Some are wet,
2.) Some are dry,
3.) Some unknown combinations.

There are infinite subsets of these general types. Any is possible.

EXISTING STANDARDS DEFINE DEBRIS, SOILS AND CONTAMINATION...

1.) In two dimensional "diameter".

2.) In relation of "distance" contamination-to-core.

3.) There is o consideration of 'connector geometry' such as the alignment sleeve, or, adapter housing.

4.) Existing Standards are based on 'easy-to-remove' debris

Myth or Reality?

There is "SCIENCE" to:

The various types of contamination, how to remove this contamination, and, how to actually see surfaces that need to be cleaned.
So, let's do a deeper dive…

What is "dry debris"?

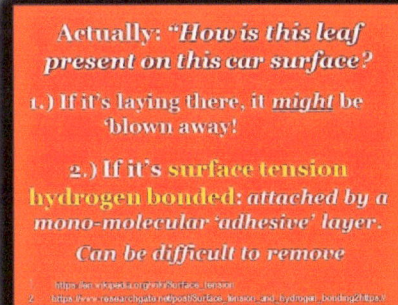

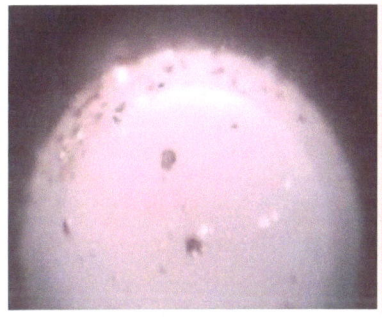

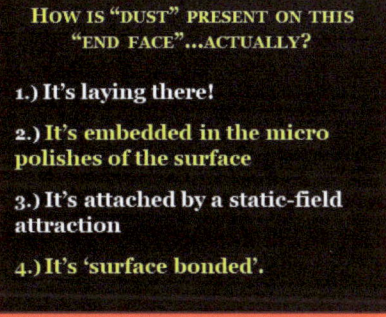

While the vehicle surface and the end face are <u>not</u> the same … there <u>are</u> common sciences of physics and contamination, as well as removal, that apply to both.

THIS IS "THE SCIENCE OF CLEANING"

(Understanding the nature of debris helps you do your best work)

If you are interested in learning more, please let me know …

What is "fluidic contamination©"?

THIS IS CONDENSATION CONTAMINATION.
(Obviously...something wet!!)

This can occurs when there is a temperature differential: the jumper in your truck in the Winter is installed in a Data Center at a constant 72F.

Notice how the fluid is evident and extends outside the 'field of view' of this 400x inspection device.

(red arrows)

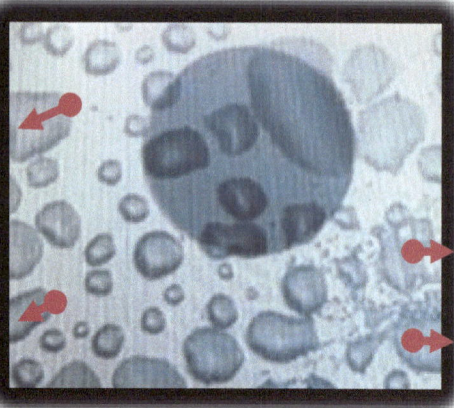

PERSPIRATION COVERS THIS COMPLETE END FACE

← Notice the three-dimensional nature of the connector as viewed through 'virtual 3D' imagery.

Fluid exists on the 'horizontal end face' and can also be present throughout other sectors of the connector geometry.

The red arrows define dry debris on the 'vertical end face/ferrule'.
It's an area not commonly seen or considered when precision cleaning.

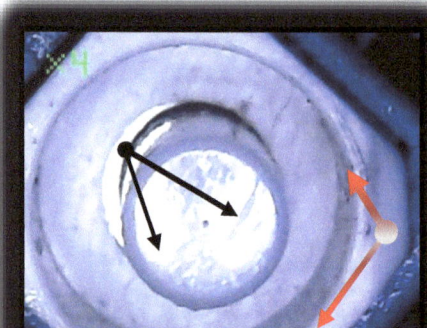

What is the difference between something that is "dry" and something that is a "fluid"?

- ✓ A "dry contaminate" tends to stay in place.
  It may not be easy to remove.

- ✓ A "fluidic contaminate", by nature, moves and transfers.
  It can flow to other sectors of the connector 'geometry'.

These "Sciences of Contamination" have been studied for thousands of years...and now we apply them to fiber optic precision cleaning and inspection. These methods and procedures have been used by your contemporaries for nearly 20 years. In many ways this is common-sense. However, as Voltaire says:

> Common sense is not so common.
> — VOLTAIRE

Another type of contamination is "damage"

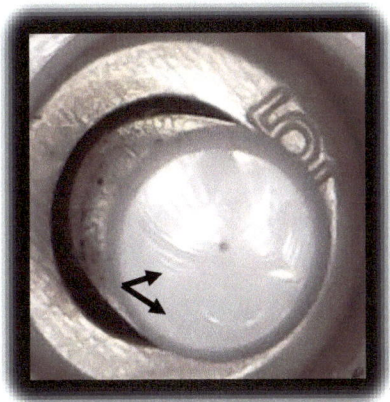

Some "scratches" are inevitable and acceptable.

Others, such as these, can trap contamination and make removal difficult. *At these microscopic levels, a scratch can be a 'reservoir' for debris.*

This understanding directs us to the significance of the interaction between inspection, location of debris, and, removal procedures.

There are numerous 'soil points' not recognized and therefore not characterized and trained. The current thought processes are to 're-clean'. The 'best practice' is to improve cleaning and inspection procedures so that multiple cleaning attempts are not necessary. The goal is First Time Cleaning.

The 'smiley-face' gouge extends outside the limits of existing standards. The general concern would be whether the defect would impede transmission by nature of the scratch itself. Viewing this connector through digital photography provides a 'virtual 3-D' perspective that depicts this defect as a possible reservoir for debris.

Awareness to all of these potentials is knowledge of these possibilities. This may be 'actionable' by cleaning or replacement...or it may not. At the very least: you know!

The 3rd Type of Debris
"Dry Types and Fluidic© Types" in "Combination"

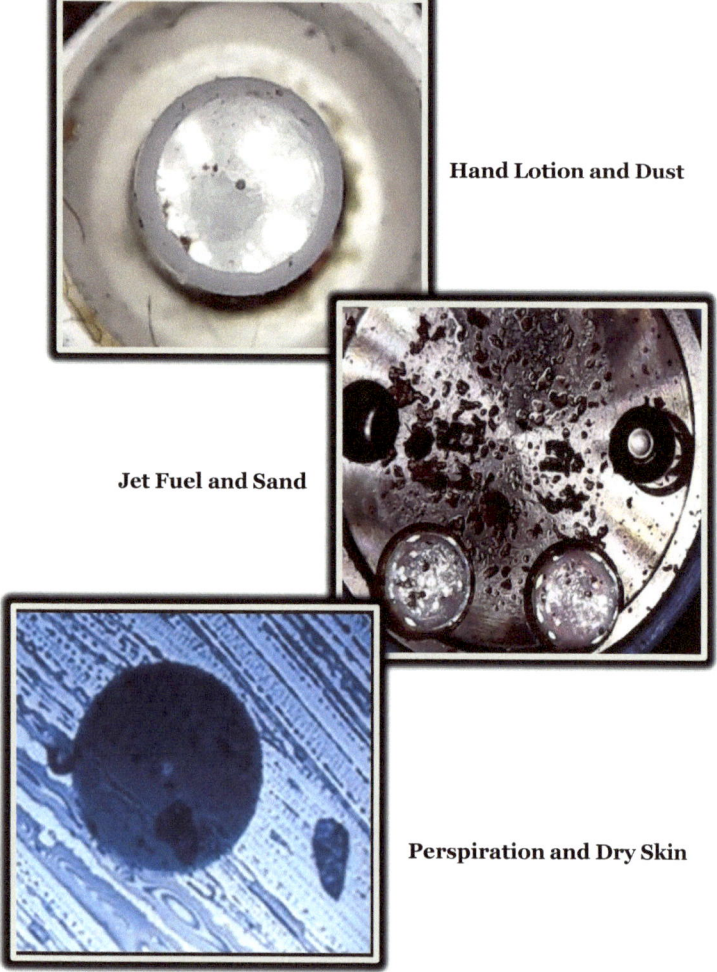

Hand Lotion and Dust

Jet Fuel and Sand

Perspiration and Dry Skin

"Combination Contamination" may be the most common. It also can be the most difficult to remove as they can be mixtures of ionic and non-ionic types. I would not imagine these mixtures are ever '50-50' but rather infinite quantities of various types. Both the Cisco® Series and my follow-up tests in 2014 studied 'combination contamination'. Combinations of contamination are not mentioned in any standards.

Training of any cleaning procedure should include a discussion of the three general types of debris. As well, as you will read, there are limitations to existing cleaning 'specs' and 'products' that can be easily remedied. Proper inspection and cleaning does not mean you have to spend more. First time cleaning should cost less!

PRESENTED ARE ONLY A FEW OF THE (INFINITE) TYPES OF (POTENTIAL) DEBRIS THAT CAN INFLUENCE INSERTION LOSS
-Also remember that 'insertion loss' is the technical name for an 'unhappy client'-

- "The Problem" is that existing standards are based on 'easy to remove debris' ... not real world types likely you encounter.
- Existing standards only consider a small sector of the overall connector surface ... another 'concern'.
- Cleaning tools are commercialized on 'easy-to-remove (minimum) requirements' not "best practice/future proof" (real world/worst case).

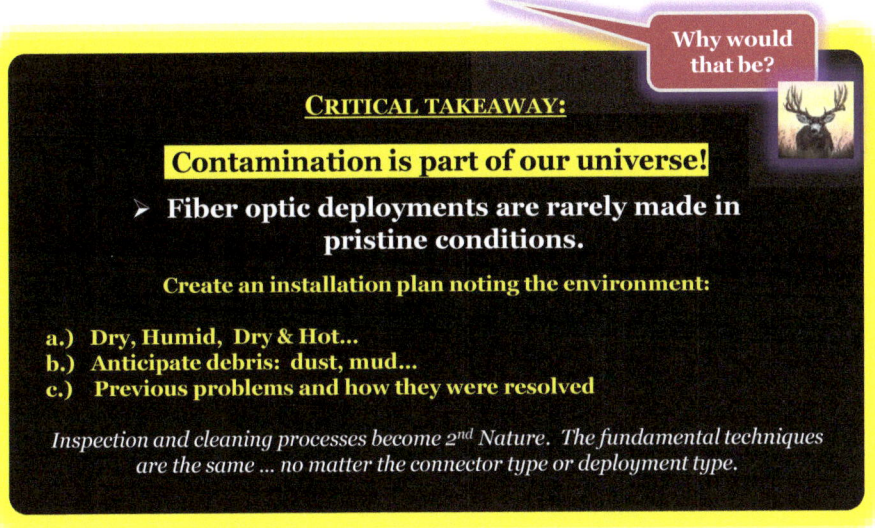

CRITICAL TAKEAWAY:

Contamination is part of our universe!

- **Fiber optic deployments are rarely made in pristine conditions.**

Create an installation plan noting the environment:

a.) Dry, Humid, Dry & Hot...
b.) Anticipate debris: dust, mud...
c.) Previous problems and how they were resolved

Inspection and cleaning processes become 2nd Nature. The fundamental techniques are the same ... no matter the connector type or deployment type.

Why would that be?

Clean & Good to Go? *How do ya' know?*

The global rapid expansion of fiber optic networks and installations is one of the unsung and understated advancements of all time. "History Channel" will look to the last 30 years in amazement of how much has been achieved and how the sciences have advanced.

At this point in time, there are two ways, with many subsets, of how we, as an Industry, disseminate our knowledge. One of them is through 'standards'. The other is by those manufacturers who invest in research and development to bring new products to the stage. Some of these products are 'worthy' and others 'not so much'! Deciding which, in an unbiased 'vendor neutral' stance, is not often easy.

At times these decisions are purely 'financial'. Other times the decisions are made by personal recommendations and points-of-view. At times, real science is clouded by relationships.

So, what works best? What criteria can you use to make an 'informed decision' without these filters? Let's see if discussion is possible for fiber optic precision cleaning and precision inspection.

How do to know if a connection is 'clean' and 'good to go'?

By Standards

IEC 61300-3-35
- TIA 455-240
- Telcordia GR 2923-CORE
- SAE AIR-6021

By Testing

OTDR
- Power Meter
- Visual Fault Locator
- Direct or Video Inspection

Can Standards be obsolete and Testing Inadequate?

Yes

Remember the only way to know if a connector is clean is to see it.

If you are not visually inspecting...or, using the incorrect testing tools...you are improperly testing!

For Fiber Optic Sciences, 'the 5-10 year update cycle of standards' for cleaning and inspection does not keep pace with development of transmission technologies.

Anatomy of a Standard
Look outside our industry to understand the concepts

SAE J 429 Bolt Standard:

Grade	Size Range	Tensile Strength min PSI	Minimum elongation %-	Material Hardness
1	¼"-1 ½"	36,000	18%	B7 to B100**
2	¼"-1 ½"	57,000*	18%	B80 to B100**
5	¼"-1 ½"	92,000*	14%	C19 to C30
8	¼"-1 ½"	130,000-	12%	C33 to C39

I chose SAE J 429 because most of you are aware of it, or, the NAS Standard for Aerospace.

Grade	Possible Application
1	Toy Assembly
2	Light weight lawn mower assemblies
5	Medium weight assembly such as riding lawn mower
8	Heavy Duty assembly: bridges, vehicle suspension

SAE J-429 IS 'APPLICATIONS SPECIFIC'.

It was first published in 1983 and updated in 1999. Not much changes with "nuts and bolts" standards like this! (The 1999 Update added "washers"!)

Fiber Optic Inspection and Cleaning Standards are considered based on 'easy-to-remove' debris. Based on that concept, it's possible you would be faced with a "Grade-8 Problem" and a "Grade-1 Solution"! Fiber Optic inspection and cleaning standards should be based on 'worst case', applications specific concepts. When we designed and competed wheel-to-wheel, it was always with the consideration of "worst case leads to best practice".

> **Rapidly Evolving Technologies** (such as fiber optics) **do not wait five or ten years to 're-standardize'!**
>
> As it is now: Rapidly Evolving Technologies evolve based on Proprietary know-how, Trial-and-Error, White Papers, Instructions, Word-of-Mouth Training -versus- Formal Training (sessions such as this).
>
> ➤ FIBER OPTIC STANDARDS CAN BE INTERNET BASED, BLOGGED AND VETTED BY SME FOR ANNUAL UPDATE.
>
> ▪ I advocate the concept of a **Rapidly Evolving Internet-Based Standard.**
>
> ▪ Input from global social media: formal vetting at international conferences that many can't attend.

Should we:

Eliminate Standards?

Mini Conclusion:

Eliminate Standards?

ABSOLUTELY NOT!

Think in other ways:

Understand that existing standards are 'minimum requirements' and may not be: 'Best Practice'.

Create your own:
'BEST PRACTICE STANDARD'

Make it part of a Work Order or Network Design

Update cleaning and inspection procedures using the model:
"Worst Case Leads to Best Practice".

How to create your own "Best Practice Standard"

1.) Document similar operations: create a log.

2.) Include cleaning and inspection in your pre-work and post-work briefings

3.) Review and update prior to 'next job'

A pattern will emerge that generate 2nd nature, habitual cleaning & inspection techniques.

THIS IS GOOD TIME TO REMIND YOU OF TWO VITAL FACTORS:

1. **Precision Cleaning a Fiber Optic Surface is "Applications Specific"**
   - The type of debris, location of the job, tools you select all play a critical role.
     - *The process is not 'one-size-fits-all'.*

   ➢ When I clean, test and evaluate: I seek 1st Time Cleaning

2. **End Face and Connector cleaning is NOT the same as _Fusion Splice Prep_**
   - The Products may be different
   - The Techniques may be different
   - Fusion splice prep cleans jacket residues and the SIDE of the FIBER… _not the end face_

"Worst Case" leads to a "Best Practice Standard" you can implement… this afternoon!

Fiber Optic deployments are: in clean(er) data centers, along a rail road right-of-way, deep in an arid military theatre, during a rain-storm, in a bucket-truck, in Kansas, Ship Board, or Aerospace … FTTh in Ohio … an MDU in The Bronx … FTTp in Tampa!

Always remember…no matter your specific discipline…the fiber core is the name of the game and it doesn't know where it is!!!

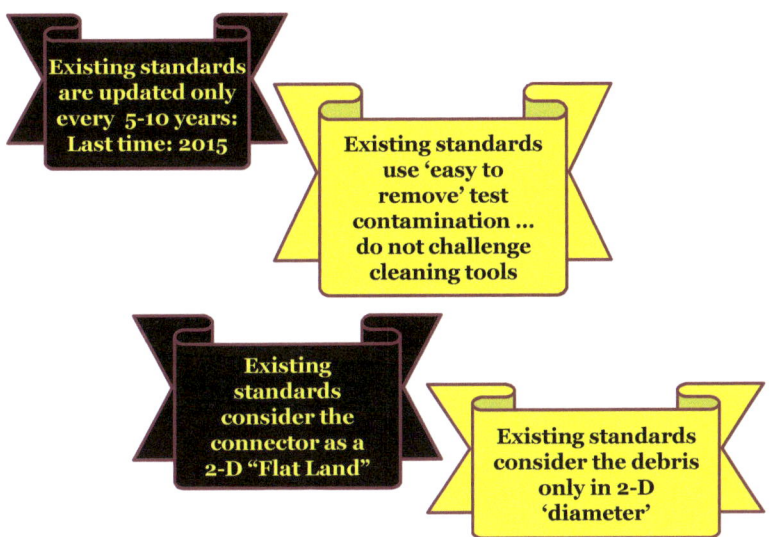

MINI CONCLUSIONS

- Existing standards are updated only every 5-10 years: Last time: 2015
- Existing standards use 'easy to remove' test contamination ... do not challenge cleaning tools
- Existing standards consider the connector as a 2-D "Flat Land"
- Existing standards consider the debris only in 2-D 'diameter'

Eliminate Standards?

- Obtain support from the cleaning tool producer or manufacturer's representative.
  - Your 'distributor' is a critical *supply source* ... but may not be the best resource for this information.
    - Give me a call...

Absolutely Not... redefine them

WHAT DOES IT ALL MEAN?

Connector Types...
Debris Types...
Standards...

> **What is Precision Cleaning ...**
> **Precision Inspection ?**

A Higher Standard to "Future-Proof" your success...and establish "Best Practice".

This is necessary because existing methods and procedures are at least 30 years old.

Prior to 1990, the global chemical inventory was based on chlorinated solvents. You may recall the 'push' to eliminate CFC's. At that time when 'CFCs' were used to clean, IPA was also a 'favorite'. As CFC's were gradually phased out, by the end of 1995, the new chemicals had not been completely formulated. This reality opened the door to "Reagent Grade IPA" as the default cleaner. By the early part of the 2nd Millennium, the new non-CFC solvents were formulated and introduced. Many of them are used today in popular branded products specifically formed for the fiber optic industry.

Simultaneously, the fiber optic industry was challenged to develop new and better inspection. At first these devices were mini-TV cameras attached to a monitor. These products, spear-headed by Westover Scientific, were cumbersome instruments that evolved to 'convenience instruments' ... which you may use.

However, the 'weak links' were that these devices compromised magnification for resolution. This meant that optical companies used higher magnification to see debris better. Doing this meant less of the surface was seen. 100x grew to 200x and soon 400x became what 'everyone' was told they needed! Had there been a high-resolution 100x scope that saw debris in better definition, there would have been fewer problems associated with cleaning! As well, the cameras were limited in their quality and this become more of an 'issue' with price pressure. Based on these limitations, cleaning and inspection standards were first developed in 1998 and continue, largely unchanged and updated to this time. Existing microscopy is based on existing standards and only sees a very limited section of the end face on the 'horizontal surface'.

Since fiber optic deployments are dependent on pristine clean surfaces and since deployments are in such varied environs, a higher standard is desirable that 'sees' more of the connector beyond the limited horizontal surface.

"Location ... Location ... Location!"

THE FIVE ZONE THREE DIMENSIONAL VIEW

The "Field of View" *within the black box* is the typical view of a 400x video scope. It's limited to a radius of ~250-300μm from the "core".

- **The greater the magnification** the better the resolution (ability to define contamination)

  → but ←

- **The greater the magnification, the <u>less</u> is** seen *of the complete connector surface.*

- To scientifically precision clean we must precision inspect. The means to not only look outside the box, but also, think outside the box to this higher standard.

- If cleaning is important to successful deployments…it's time to update our understanding to three-dimensions.

- This means add "Zone-4" to complete the views of the horizontal surface and include "Zone-5" to consider all potential contamination points.

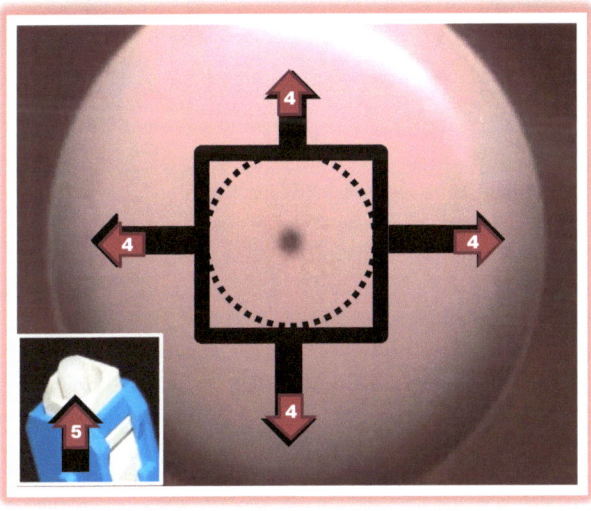

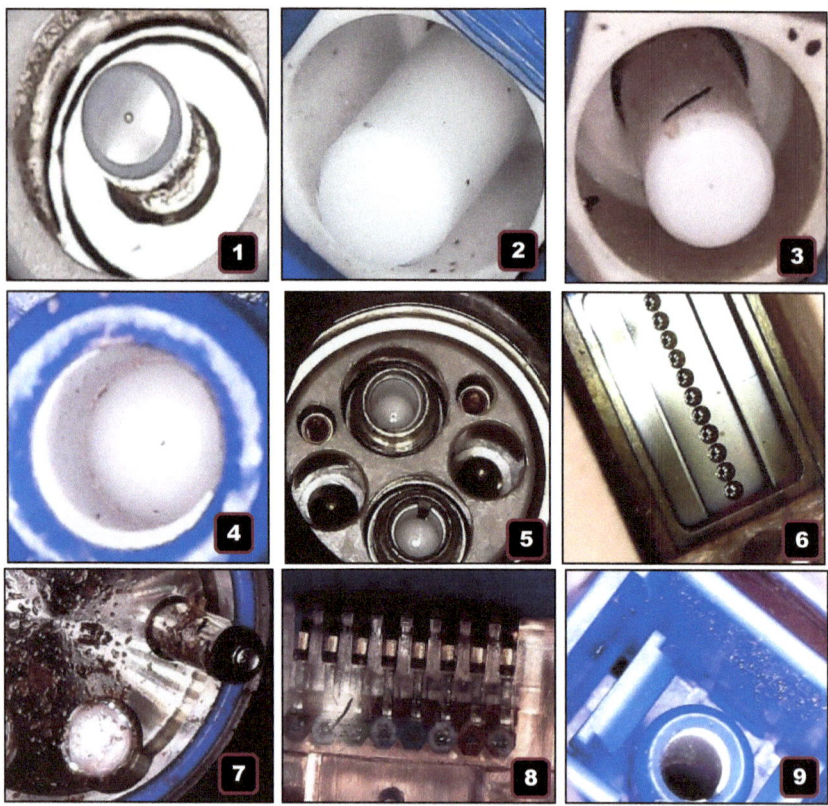

To properly precision clean and inspect...
"Look Outside the Box"
➢ Consider total 'connector geometry'

Why 'outside the box'?
Debris can be located in places you can't see.

1.) Clean end face; contaminated Zone-5
2.) Clean end face; contaminated ferrule surfaces
3.) Another 'lucky picture' of unusual debris on Zone-5
4.) Picture looking at debris on adapter; Zone 1-2-3 clean
5.) SMPTE LEMO® hybrid connector with single mode, high and low voltage
6.) Amphenol MT Expanded Beam Surfaces
7.) Debris on TE Expanded beam pins and inter-surfaces
8.) When was the last time you viewed a RG Pins for alignment?
9.) Dusty adapter and who among us has not (inadvertently) touched an adapter housing 'en route' to the alignment sleeve!

Images captured on RMS-1™
Rotating Adapter© Video
Inspection. (Patent Pending)

Precision Inspection:
The Concept of "Field of View"

One of the seminal events in my career happened in about 2002 when I visited Glen Porter, founder of MicroEnterprises and the OptiSpec® Series of video inspection devices.

Glen Porter was one of the founders of inspection, I recall his early work on 'auto-detect/pass-fail' in the 2005-06 time frame. Many of Glen's scopes employed a 'scroller' which enables the video camera to scan 'east-to-west; north-to-south' outside of the field of view of 'fixed view' instruments. For many years I demonstrated cleaning and inspection on his ME-9500. This is a video was taken of an end face using the ME5000 with a "scroller" that enables inspection of the compete field of view. Porter's work inspired my development of RMS-1 using digital photography.

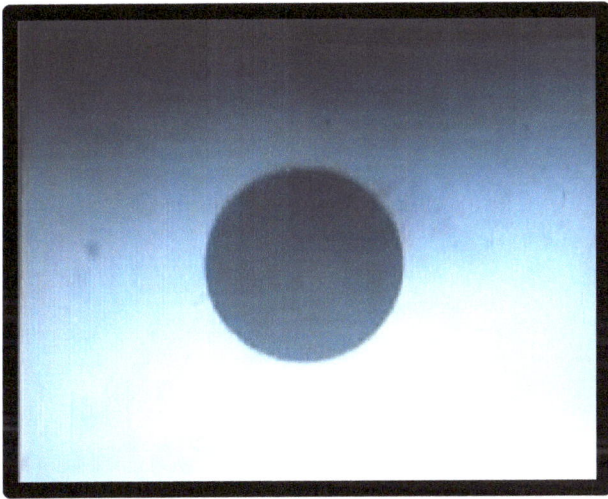

This video, along with several others in this book, are available at "EdForrest" on YouTube®.

If you are enjoying the MP4 version of this work...the video should be playing now!

This is about half-way!

Please, take a short break. Jot down your thoughts, notes or, challenges and send Ed an email.

edforrest@fiberopticprecisioncleaning.com

Welcome Back!
(If you took a break!)

Questions or Challenges?

Please remember I learned what I know from you! Visits to installation garages, formal approvals, and, trade shows all influenced these studies. I "depend" on your input: let me know your thoughts, successes, and, experiences.

edforrest@fiberopticprecisioncleaning.com. +770-971-8100 USA

Don't forget! There are video on YouTube®. Search "Ed Forrest".

The Three Dimensions of Cleaning Anything
INCLUDING ALL FIBER OPTIC CONNECTIONS

"Cleaning" is one of those second-nature events! Washing a car, dishes, clothing or a floor are tasks that we don't think too much about: we just do it! However, before we start we 'automatically consider" these three things.

My experience is that this is not often the case for critical fiber optic surfaces.

While we may not always think about it this way, there are three things we consider before we clean...anything!

1.) What soil am I cleaning?
2.) Where is the soil?
3.) How will I remove it?

An imbedded & integrated aspect of *"The 5,000 year Cleaning Heritage of Humanity"* we have an advantage:

We can see what we are cleaning!

This is not always the case when inspecting fiber optic surfaces...

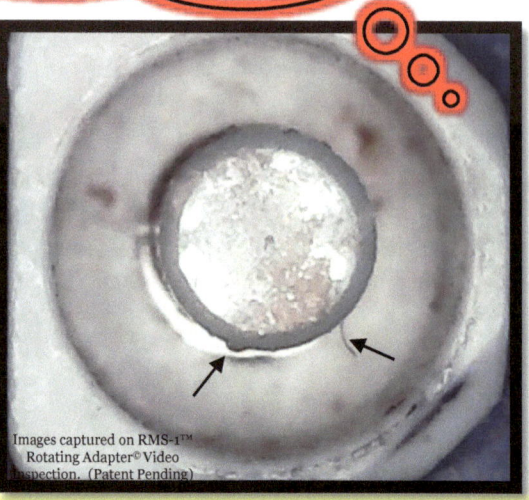

Images captured on RMS-1™ Rotating Adapter© Video Inspection. (Patent Pending)

There are two topics!

The first is that existing inspection, as you know it, can't see this range of debris on these surfaces.

Notice the surface at 6:35 location on the 'vertical' surface. There is debris 'standing proud'. Likewise @ 4:25 location there is a strand of lint or hair.

The 2nd concern is that this image 'passes' existing standards...

...even though there is transferrable debris on the "vertical ferrule" and other "inter-surfaces". There are times when we can't see what we need to see!

This is a video taken with RMS-1. If you would like to have a copy, please let me know. If you would like to demo RMS-1, please see the "virtual demonstration".

You Tube® → https://youtu.be/dI-Djpnzd4Y
Vimeo® → https://vimeo.com/208054841

The Myths of Cleaning

The reality is:

...you may not be able to see what "needs to be cleaned" on a fiber optic surface.

This session is intended to increase your understanding of the dynamic interaction of debris type and location.

Some surfaces may need to be cleaned and others not.

All should be considered.

How do you clean a fiber optic surface?

These are the cleaning methods available in 2018.

1. **Dry Cleaning**
2. **Wet-to-Dry Cleaning**
3. **Hybrid/Combination Process**
4. **Blind-Cleaning**© "leap-of-faith"

AUDIENCE PARTICIPATION:

Which technique do you use? If you would like to help me compile a statistic, please send me an email!

Which one do you use?
1.) _____
2.) _____
3.) _____
4.) _____

Existing standards, and likely the way you were trained, only mention the first two.
- The 3rd technique is recorded as a Telcordia GR.
- The 4th technique is "blind cleaning"©... without inspection.

Surprisingly? My survey result, conducted over a five year period, polling more than 3,000 users: ***60% of all connections are "blind cleaned".***

Sure, you should purchase a video inspection device...but...I also understand this may not be possible. This book and seminar are intended to encourage you to buy the equipment you need...and also. Improve your procedures as a "safety net" when you can't!

WHAT DO YOU USE?
FIBER OPTIC CLEANING PRODUCTS: 1998-2018

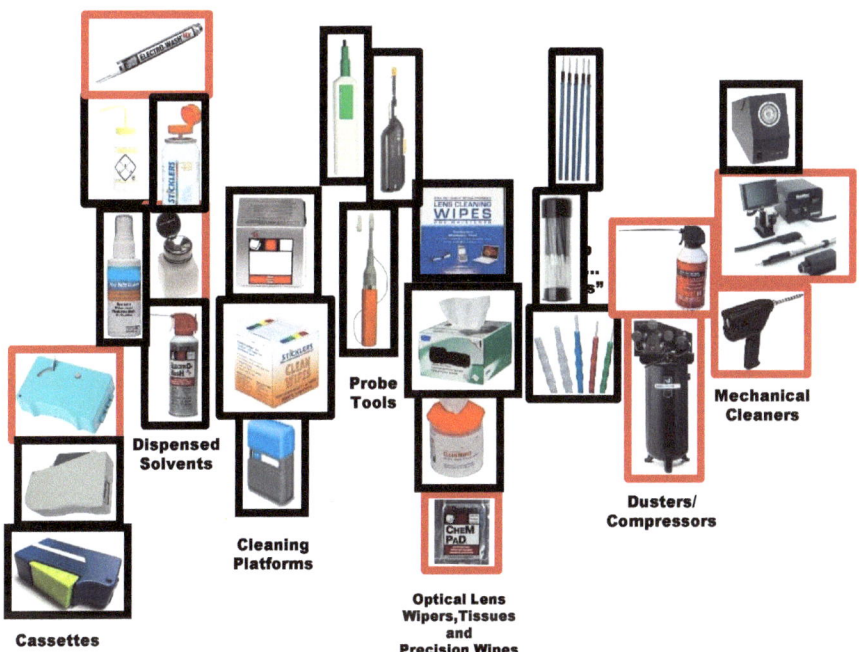

As you see, there is no limit to the selection of fiber cleaning products! It's not unlike selecting a laundry detergent: each product has someone who designed the product and a marketing department that sells it to you!

Let's study the ways and means you were taught to clean a fiber optic connection. The current thought is to begin with the "dry method" and *if that does not work out for you* to use the "wet-to-dry" technique.

Let's take a look. The base of this study began in 2001 with repeated tests since that time. One of the fundamental tenets of science is the ability to provide a repeated result. What you are about to see has been repeated more than 10,000 times!

How is this possible? I looked at the number of garage training sessions, seminars and trade shows, leads from demonstrations, and, arrived at 10,000. It's s conservative estimate. (There are hundreds of admissions badges handing in my office!)

"How Do You Clean?"
Dry Cleaning Process
Using tools, swabs, and wiping materials

ADVANTAGE: "THE CONVENIENCE OF TOOLS"

Perhaps the very first cleaning device was the NTT Cletop® I first saw the 'reel cleaner' along with 2.5mm swabs at Pirelli Electronics. At this point in my business life I was more involved in "selling"...even then I knew this was formidable competition. Nearly all the fiber-techs walked about the work floor with one in their shirt pocket or jeans. As the years have passed, other companies cloned the tool and even NTT did a "Chevy-Buick" thing by introducing their own competition.

To this day, many consider CleTop® and OptiPop® as the standard tool of this type. These are dry cleaning tools with a rather limited 1" long cleaning surface. While they can be moistened, there is little room for a 'drying procedure" unless another wiper is presented in the cleaning window. At one point in time, each 'click" was a handsome cost of $0.75. That didn't matter, This decision is about convenience.

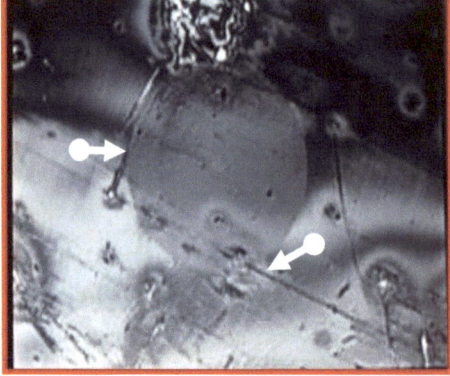

Disadvantages: The image above is actually a video posted on YouTube®. This is also a repeat demonstration performed many thousands of times where attempts to remove a complex soil resulted in the smearing you see above. It was from these demonstrations and images on the ME-9500 that I began to observe the 'top' of debris would be moved while the base would remain. Above the 'white arrows' point to this reality about "dry cleaning": the process moves debris and may not remove it.

There is more. In 2006 a lab test was performed that generated a rather low static field that attracted an excessive amount of debris to the end face. Dry cleaning can generate static field contamination.

In recent times another fundamental about 'dry cleaning' came to the forefront: "dry cleaning" works best for contamination that is wet. It's not as much a cleaning action as it is a 'mopping action".

When you think about it, this is not a surprise! Would you try to remove mud from your car using a dry wiper? How about cleaning an LCD screen with a dry cloth that only attracts more dust! The coffee spill...you used a dry wiper to control that contamination.

> **Clue: 'What' you use to clean is not always the same as 'How' you use it!**

Dry Cleaning Process
Using tools, swabs, and wiping materials

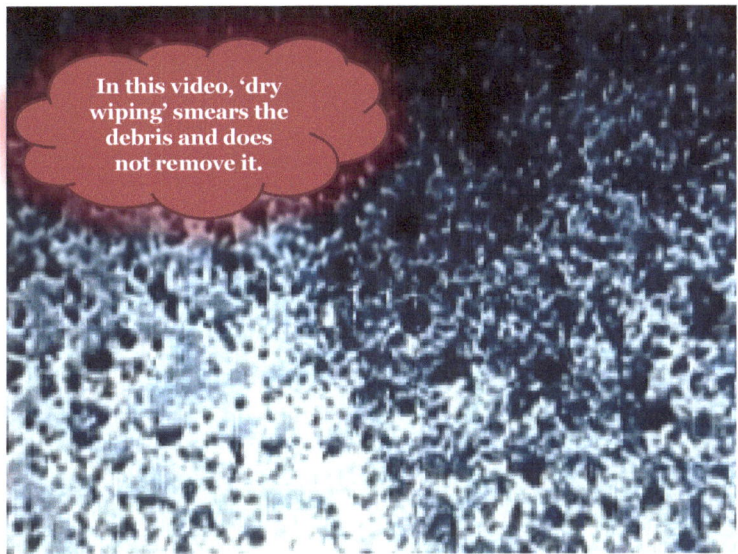

If you have a copy of the MP4, this video plays automatically.

Otherwise, why not go to YouTube® and check "Ed Forrest" to this how the debris smears and fiber begins to appear. Look carefully to the center upper right quadrant of this picture...you'll see the barely visible fiber.

Using a tool designed to clean "dry"...this debris only seems worse. Repeated cleaning actually seems to embed the debris further to the ferrule.

Dry cleaning 1st may not be "Best Practice"

Remember the 'last' coffee you spilled on your desk?
There was a small river headed to your lap and keyboard!!!
How did you stop it ?

- You used a dry wiper: *Dry Cleaning is a "mopping action" that absorbs something "wet"*.

    - Best Practice for *Dry Cleaning* with a probe tool, swab, cassette or cleaning platform: <u>use the procedure when (if) you identify the debris as a fluid</u>.

        - If you are *not* inspecting the surface...every time...
            - how do you know?

So, (following existing training/standards),
the instruction is:
"Since dry-cleaning didn't work...
...use the wet-to-dry method."

THAT'S THE INSTRUCTION ... CORRECT?

I believe words have meaning and without definition, something as critical as precision cleaning a fiber optic surface should not be left to individual interpretation.

If you've ever snapped the head off the top of a bolt...you understand what it means to use a torque wrench!

WHAT'S IN A WORD?

Everything in Moderation!
(There are micro-surfaces)

Does everyone really understand when we say: "Wet-to-Dry Cleaning"?

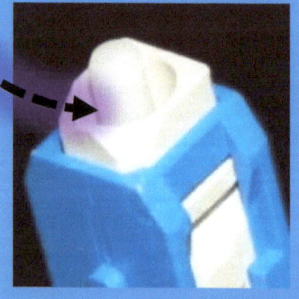

DO THE WORDS MEAN WHAT THEY SAY?

Sometimes when I get to this point someone will say that "everyone knows what it means". To be honest and real about this...there are very few absolutes and my experience is that many who clean believe they can dry!

This is simply not always true. These surfaces can be 'flooded' with too much cleaning solvent. The following demonstration, which I urge you view on YouTube, has been conducted as many times as the "dry cleaning" experiment: many thousands of repeated same results after asking technicians the question: "How do you use a wiper with solvent/"

The response is to either use a presaturated wiper or 'make one' by wetting it. The end face is then drawn through the wetted wiper...perhaps 1-3 times.

I believe that fiber optics has a long future. This means there will be many deployments and many new crafts-persons. This also means our instructions, standards and training must be as precise as the surfaces we are cleaning.

"Is there a better way?"
Is something we all should ask more often.

"Wet-to-Dry Cleaning"

- **"Do words mean what they say?"**
   "WHERE'S THE BEEF?"

Hey Ed, what's wrong with saying "wet-to-dry cleaning"?

"Wet-to-Dry" Cleaning

This demonstration is conducted using 99.9% IPA and is based on a process technicians relate as "...*the way we do it*...".

The demonstration has been repeated many thousands of times over the last 20 years.

The testing clearly proves it is possible to "flood" the end face...check it out!

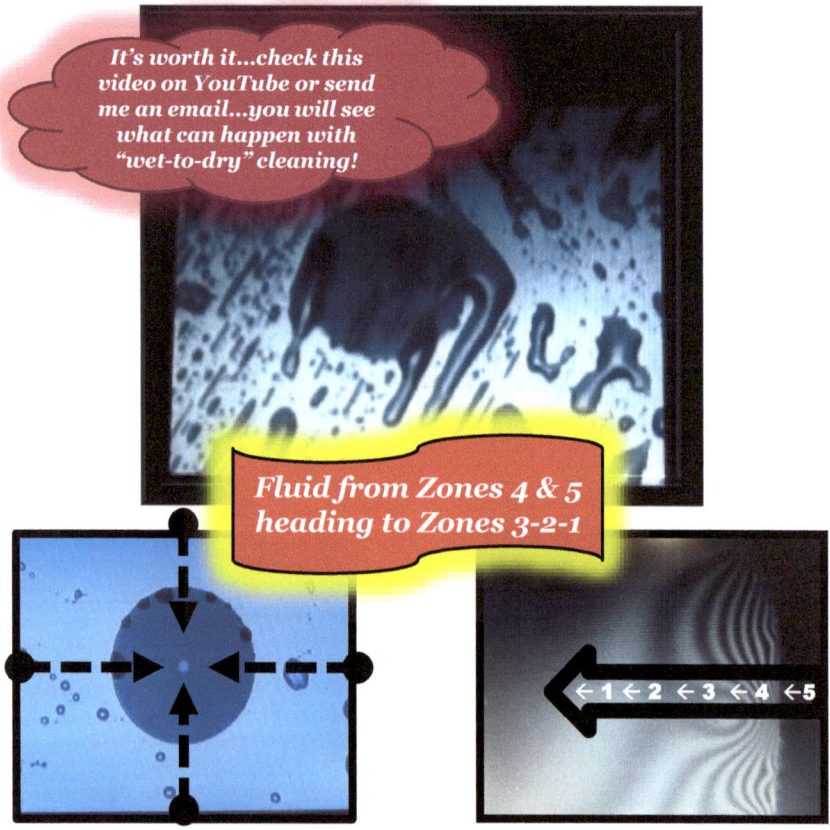

It's worth it...check this video on YouTube or send me an email...you will see what can happen with "wet-to-dry" cleaning!

Fluid from Zones 4 & 5 heading to Zones 3-2-1

Don't let the term 'wet-to-dry' convey a sense of "cleaning-over-confidence".

The Goldilocks Zone:
The 'wet-to-dry' process may, *or may not ...*
actually dry the surface.

<u>*Stop and think about it*</u>: Using something "wet" to clean works best on debris that is "dry".
This is the Science of Hydrology*: *the attraction and absorption of dry matter by fluids...and fluids into dry matter.*

*Professor Tyson Ochsner: Oklahoma State University

I spoke with Professor Ochsner and exchanged email regarding Hydrology. His science is most often associated with water and absorption. He agreed other media than water are also considered hydrological science. That science reconfirms what a colleague, Paul Blair, taught: 'soils are attracted to moisture".

In recent times, NASA's spacecraft, The Kepler Observatory has discovered what may be many thousands of planets in "The Goldilocks Zone": not too hot and not too cold to sustain life as on Earth. There is also a "Goldilocks Zone" when considering fiber optic precision cleaning and inspection: too little or an incorrect solvent and wiper combination: the debris will not be removed. Too much and the surface will be flooded: difficult to dry or problematic in the time of post cleaning and inspection.

Throughout this seminar reference is made to "science": history helps us focus the Science of Precision Cleaning and Precision Inspection for now and future times.

The selection of tools, claims and counter claims, and confusion about which works best or lasts longest has resulted in lack-luster performance. Technicians relate how they spent a half-hour cleaning, or replaced a jumper in frustration.

There is a strong commercial undercurrent to an aspect of science that is not all that difficult to understand! Cleaning products themselves are often termed "disposables". This means repeat sales: think about that the next time you select a product.

'Fluidic Contamination' on the outer limits
of the 'field of view', and embedded in Zone-5,
can transfer and migrate.

If you are "blind-cleaning" you will never know!

> Suggestion: *If you are "wet-to-dry cleaning", scope after cleaning and <u>then again</u> in 3-4-5 minutes to assure fluid has not transferred to the core.*

Food For Thought and Mini Conclusions

"Wet-to-Dry Cleaning"...works best for
<u>'DRY DEBRIS'</u>.

1.) Dust on an LCD Screen

2.) Dried mud on a painted surface

3.) Dry Debris on a fiber optic surface that may be excited to attract more debris by static field contamination.

- Dry Cleaning works best for 'wet contamination' ... *it's a mopping action*

  - What's the "Best Practice"?
  - What works on the widest range of debris?
  - Can a (different) 3rd PROCESS clean The 1st Time?

> FOOD FOR THOUGHT:

IT'S CONCEIVABLE THE INSTRUCTION: "dry clean first" and follow with "wet-to-dry cleaning" is:

- Counter-productive
- **Reverse** Best Practice, applications-specific precision cleaning procedures.
- Does not look to future needs.

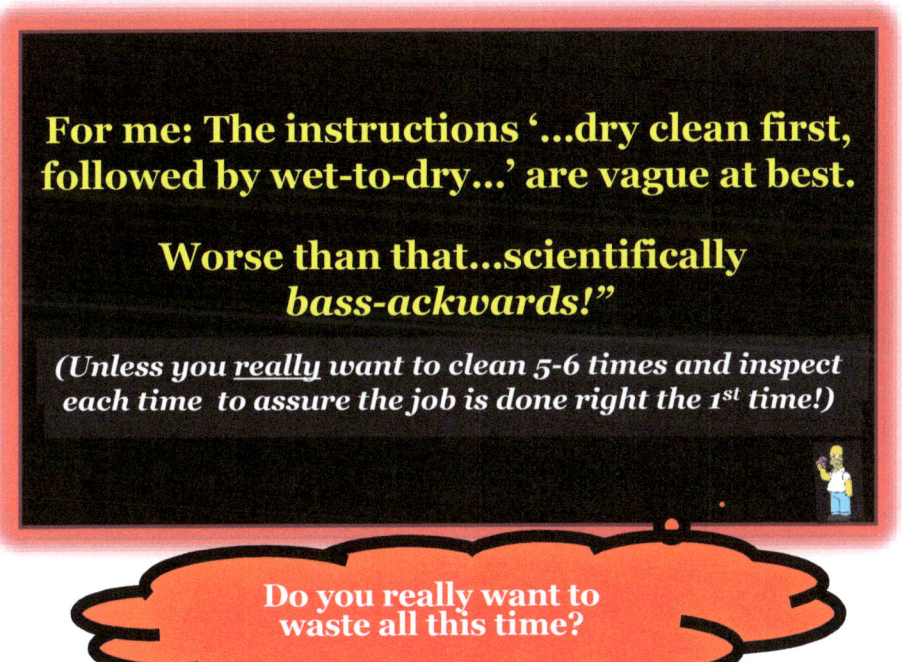

For me: The instructions '…dry clean first, followed by wet-to-dry…' are vague at best.

Worse than that…scientifically *bass-ackwards!*"

(Unless you really want to clean 5-6 times and inspect each time to assure the job is done right the 1st time!)

Do you really want to waste all this time?

The 3rd Process

- Recorded in Telcordia GR-2923-Core: 2011

✓ Identifies proper cleaning materials

✓ Identifies a technique that <u>under-uses</u> high-performance fiber optic cleaners

✓ Identifies and Integrates and a drying technique

✓ Applicable for all current cleaning tools and enhances their performance: *<u>more often than not,</u> this 3rd procedure makes the product you select perform better.*

The 3rd Process

THE 3RD TECHNIQUE USES A SMALL AMOUNT OF HIGH PERFORMING FIBER OPTIC CLEANER.

This assures that the end face is not "flooded" with excessive cleaner that can be trapped in the "Zone-5" recesses of the connector.

There is an integrated drying step...this means different techniques for various tools.

"Drying is essential" as an integral step to *actual debris removal from the end face.*

Yes, this may mean you "click it" more than once and you use two swabs instead of one!

- The technique is 'Plan-B' when video inspection is not available...or, you've cleaned it for 15 minutes and the *"darned thing still is not clean"* !
- At least the job is done. You can move on to the things that are really important.

THE 3ʳᴰ TECHNIQUE USES A SMALL AMOUNT OF HIGH PERFORMING FIBER OPTIC CLEANER.

This assures that the end face is not "flooded" with excessive cleaner that can be trapped in the recesses of the connector...including Zone-5

Also worth the effort. The total 'horizontal surface' appears.

The 3ʳᵈ Technique is descriptively simple. "The Point" is to use a fiber optic cleaning solvent every time. This assures if you are "blind cleaning" you have a safety net.

On the following pages are Methods and Procedures. If you are viewing the MP4 version, the video above is playing live. It is also available on YouTube. This procedure has been repeated many thousands of times, and has formal approval numbers (SSI/PIDS/PeopleSoft/CIFA).

The procedure is "field proven" since 2004.

Now that you have seen "Why", continue to understand "How"

Did you ever wonder?

"Why" are there so many different cleaning products?

Seems like 'everyone' is selling something to clean....must be great business! There is little doubt about the 'commercial success' of cleaning products! Each one vies for a market space and new products claiming to be as good as the other...are enticingly less expensive!

WHAT DO YOU USE?
FIBER OPTIC CLEANING PRODUCTS: 1998-2018

This is "Fiber Optic Cleaning Product Confusion"!

"What if..."

...the ways and means you were taught to clean a fiber optic connection. The current thought is to begin with the "dry method" and if that does not work out for you to use the "wet-to-dry" technique.

Let's take a look. The base of this study began in 2001 with repeated tests over a 20 year period. One of the fundamental tenets of science is the ability to provide a repeated result. What you are about to see has been repeated more than 10,000 times!

How is this possible? I looked at the number of garage training sessions, seminars and trade shows, leads from demonstrations, and, arrived at 10,000...as a conservative number!!!

IEC 61300-3-35 enables the technician to clean 'up to 5 times'

➢ <u>**What if**</u> ... *all of these products could work better?*

➢ <u>**What if**</u> ... *all* of these tools could work closer to The 1st Time ... every time!

"WHAT IF"

A (slightly-different) procedure served as a "safety net" in those times when the only method and procedure was to "Blind Clean"?

WHAT IF"
that (slightly-different) procedure was <u>common to all tools and eliminated confusion of choice?</u>

What if: "Precision Cleaning is a Process Change"?

> **NO MORE "MYTHOLOGY"!**
>
> **This is "The Science of Cleaning"**
>
> **These principles are applied every day in all facets of cleaning and inspection.**
>
> **Update to these precepts to improve your work in this Industry.**
>
> **Design-Install-Train-Manufacturing.**

This is how this "rolls"! No more product-of-the-month with improbable marketing promises. Let's precision clean every time and precision inspect when we can. Some will be critical of the last sentence: the reality is that not everyone can afford the 'right test gear' all the time and 100% of anything is a lofty, but sometimes unattainable goal.

By re-thinking and re-training we can have a better result because the actual cleaning procedure is improved. Rather than clean and re-clean, install a new jumper when the old one simply required a better cleaning procedure, or, sending that $10,000 circuit card to "warranty" when all it needed was proper cleaning...let's consider 'right-the-first time'.

It's going to require a little effort to change to a "Best Practice" model! You, as an installer, have it easy: you can change this afternoon! Trainers have it more difficult because this means you will have to modify your lesson plan. I believe the change to Best Practice begins at 'system/network design'. Here, consideration to the actual job site, types of anticipated debris, and even the time of year should be designed into 'the spec'. Many times when I review installation instructions, the manufacturer's guidance is to 'clean', or 'clean as necessary' without consideration of the specific nature of the equipment. There are some manufacturers who claim that cleaning isn't necessary because it was done on the production line! I am not sure about this: "Myth Happens'!

Installers: invite your local rep, have him bring two-dozen donuts, conduct an initial update and invite them back for a follow-up in 90 days. "What worked; not so much." Change your culture!

Trainers: Let's talk about it. There are reputable producers who will be eager to update your PowerPoints! If you are a BICSI®, ETA®, FOA®, The Light Brigade®, FIS® or trained by an other group, ask them to get in contact with me, please!

Designers: You have the most difficult road as writing a spec for cleaning and inspection does not fall into your normal work. Break the mold and include this essential element that may have been causing you 'pain': soiled or improperly cleaned fiber optic surfaces blamed on your design!

Manufacturers: it's not easy to re-write an instruction manual. You have access to White Papers, Web Meetings and other information vehicles. You can reduce your warranty claims by providing clear and concise installation instructions that include defined cleaning procedures for your equipment in applications specific terms.

Why is First Time Cleaning Important?

There is a logical answer: "It's a time saver". Through the years technicians related 'horror stories' of being in a cramped-hole, at 02:35 and cleaning the same connection for 'twenty-five minutes'.

My response then is as it is now: "Why?" What was in that hole, or what was being used to clean that caused such unnecessary pain and suffering! From those stores my work evolved to the thought: "If these cleaning procedures that seem so inadequate could be improved, this would be Best-Practice." I set about, in true Euclidean-Tradition, not to 'feel-good', but rather, to improve what had been to what you are going to understand now.

What is the "proof"?

Somewhere about 2004 or 2005 I became aware of "The Cisco Series"© which was a challenge to clean, in the first time, a series of very complex soils. Who could have imagined 'metal shards', 'graphite', 'dried water', 'simethicone' (infant cholic), and other very strange debris? The challenge, which may still be on The Internet, was to clean the debris the first time. The Proof was photographic evidence of ten repetitive events. It was an honor to report ten debris, ten photographs or each and only one event required 2nd time cleaning.

Inspired (again) in 2014 by this study, I conducted and published a contemporary series of lab tests using the products below contaminated with this debris. One improvement over "The Cisco Series" as well as existing standards, was the challenge of 'combined contamination'. This type of debris is not often studied…and should be because it's most likely what is encountered!

In December-2014 I conducted an updated series of tests comparing contemporary cleaning tools.

Debris and Contamination included:

1.) Vegetable Oil
2.) 10-30w SAE Engine Oil
3.) Arizona Road Dust
4.) Desert Dust from Afghanistan
5.) Jergens® hand lotion
6.) Vegetable Oil and Arizona Road Dust
7.) 10W-30 Engine Oil and Desert Dust
8.) Hand Lotion and Arizona Road Dust

The goal was "First Time Cleaning"

Results available: please see "Publications"

1. IBC® Tool
2. ClePen Tool™
3. FerruleMate®
4. FerruleMate-2
5. Sticklers® 2.5mm Swab tools
6. ITW Chemtronics® 2.5mm Swab Tools
7. QbE® Cleaning Platforms
8. Stickers® Clean Wipe Platform
9. HandiMate™ Platform

What happened?

Over 90% of the tools worked better...cleaned The 1st Time

...with the simple process change <u>not included in the manufacturer's instructions.</u>

The 2014 Lab Tests confirmed experiences (in my commercial-life) at garages, data centers, central offices, head ends, labs and development of formal internal speciation's.

The testing I performed followed the Cisco® Series protocol. Each tool was tested ten times against each debris type and recorded. The results are published: *"A Comparison Study of Precision Cleaning Methods for All Fiber Optic Connectors". www.amazon.com*

As you know, Euclid sought more proof ... as do all scientists!

In Spring-2016 I started work to prove what seems to be obvious! For nearly 20 years I had studied and lectured about the 3-D nature of connector surfaces. The 'operational word' for me is "surfaces". Many others had studied the 'end face' and those researchers helped write existing standards and some continued to develop products to meet those standards.

As I researched it became obvious that there was little enthusiasm to change the status quo. In Spring-2016, with help and encouragement from two fine persons, development began on RMS-1©. The intent of the instrument was not to compete in the 'video inspection marketplace', but rather provide the industry with new information. As of January, 2018, there are more than 3,000 digital photographs that prove what you know as you are reading this sentence: not only do we exist in 3-D, but also, connector surfaces and debris follow the same reality

> "Proof" also derived with development of a new inspection device confirming locations of debris on all fiber optic surface types.

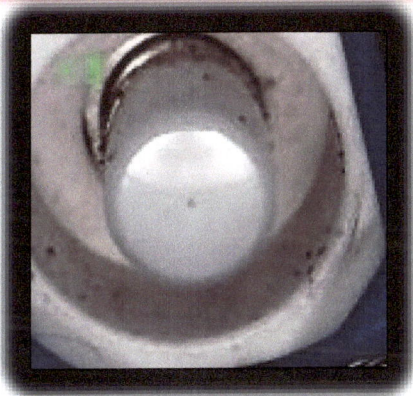

Images captured on RMS-1™ Rotating Adapter© Video Inspection. (Patent Pending)

RMS-1 has been used to study a wide range of connector and connection surfaces. Among these are common SC/LC, as well as FC and adapter housings for those types. The images of the back plane connector and housing often reveal debris never before seen.

There are a wide range of 38999 military styles that have been inspected. Some of these have been contaminated with desert dust from Middle East regions as well as Skydrol® hydraulic fluids. Evaluations of all connectors included consideration for real-world contaminants and not just limited 'easy-to-remove' types.

Included are Lemo® and Neutrik® SMPTE broadcast connectors. Images of the new Corning® OptiFit® and standard OptiFit® reveal debris on unseen surfaces that can be a 'problem child'! "MT-Types" have expanded beyond the commonplace to 'hardened connectors' such as Radiall's® Titan range.

"How to properly moisten any fiber optic cleaning tool"

Always use a small amount of high performing fiber optic cleaner. No IPA!

1.) Moisten the cleaning platform surface
2.) Lightly moisten 'reel cleaner' tape
3.) Depress the probe tool to moisten the cleaning tip.
   Don't activate ... if it "cliks" the solvent won't be in the right place!!
4.) Hold the Probe or Swab tip in the moist area for a count of 1-2-3-4-5.

Contemporary Precision Cleaning is 'all about' using less to do more.

'Best Practice' Techniques

"Lightly moisten" means to begin with minimal solvent quantity: the 'blue ovals' represent a 'solvent spot' about the size of a US-Quarter Dollar coin: 2.54cm

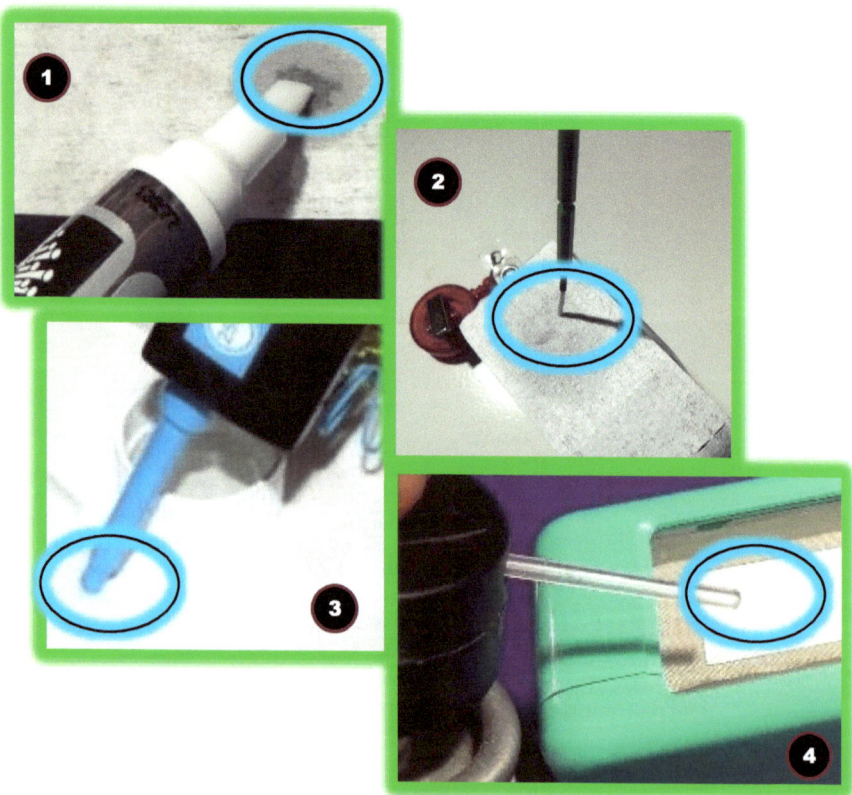

The fiber optic "precision cleaning motion" is a _straight line action._
Move debris _away_ from the initial point of contact.

This cleaning motion is easier for the cleaning platforms than reel cleaners that have smaller surfaces. Nevertheless, the process works for both cleaning tool types.

Moving debris away from the initial point of contact using a probe or swab tool is also best practice. Moisten the probe or swab tool by "solvent transfer": hold the probe tip or swab tool in the moistened area of the cleaning platform or reel cleaner…count 1-2-3-4-5. It does not take a large amount of these high quality cleaners to clean most debris. You may click 2-3 times or use a second swab tool: the first time cleans and the second dries the surface.

Some swab tips are 'flat' and others 'fold-over'. The flat tips have a smaller cleaning surface than the 'fold overs': best practice, especially if you are 'blind cleaning', is to use two swabs per connection. Swab tools have advantages to probes: your tool kit should always contain swabs.

Precision cleaning is as the decision between a 'Philips Head" and "Straight Blade" screw driver!

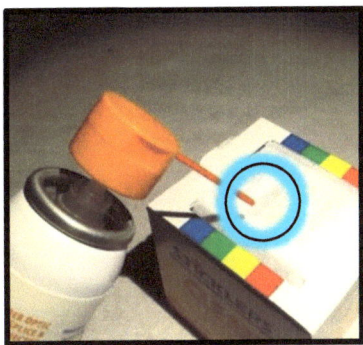 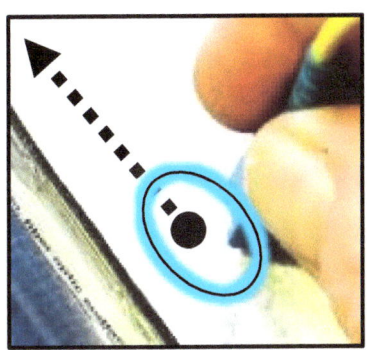

Using the cleaning platform or reel cleaner

1.) The surface is lightly moistened … ½" diameter of fiber optic cleaner is often enough!

2.) Clean from the back of the rack -or- moisten the probe or swab tool tip

A "key benefit" of the large platform tools is the surface area enables you to "find the angle". Always approach the cleaning surface at 90° perpendicular. This assures the complete end face is being cleaned.

Debris is moved away from the initial point of contact and not circled and redistributed using cleaning platforms or reel cleaners.

The Cleaning Platform provides a compliant surface for the end face. Some of integrated platens that are fresh with each tool. Most reel cleaners have hard rubber under-surfaces that are never changed. This results in cleaning or grinding debris over a hard surface rather than a gliding motion that removes contamination away from the initial point of contact.

Ask the producers to explain the advantages of their tools.

> Always be concerned and disciplined: Do not re-contaminate the surface by re-using a swab or advancing a probe only one click. Advance the wiping sheets on all cleaning platforms every after each cleaning. Some platforms provide areas for multiple single passes. Others only have one cleaning per sheet.

There are some "don't" and these are significant to successful precision cleaning results.

1.) Don't clean these surfaces in the palm of your hand or on a wiper wrapped over a finger! The intention is good as this is a soft compliant surface. However, there is potential to draw oils and debris through the wiper to the surface intended to be 'pristine clean'.

Cleaning platforms were designed to emulate a complaint surface. Some have platens that are new with each tool. Reel cleaners don't replace the rubber pads which means these become brittle. It's a good idea to replace the backing pad on your favorite reel cleaner at least once a year.

2.) The opposite of using 'you' as a compliant surface is to use a work bench or other hard surface. Remember, you are cleaning micro surfaces of micro contamination and it's likely you can damage what you are cleaning.

3.) Don't Dunk! The most complex cleaning I encountered began with the story: "We cleaned it last week. We inspected it. Now it's contaminated again." When you review the YouTube about "wet-to-dry" cleaning you'll best understand how fluids become embedded in the recesses of connectors and adapters. "Pass-Fail" gives the "OK"...but excessive solvent or debris unseen becomes the problem...the week after you leave on vacation!

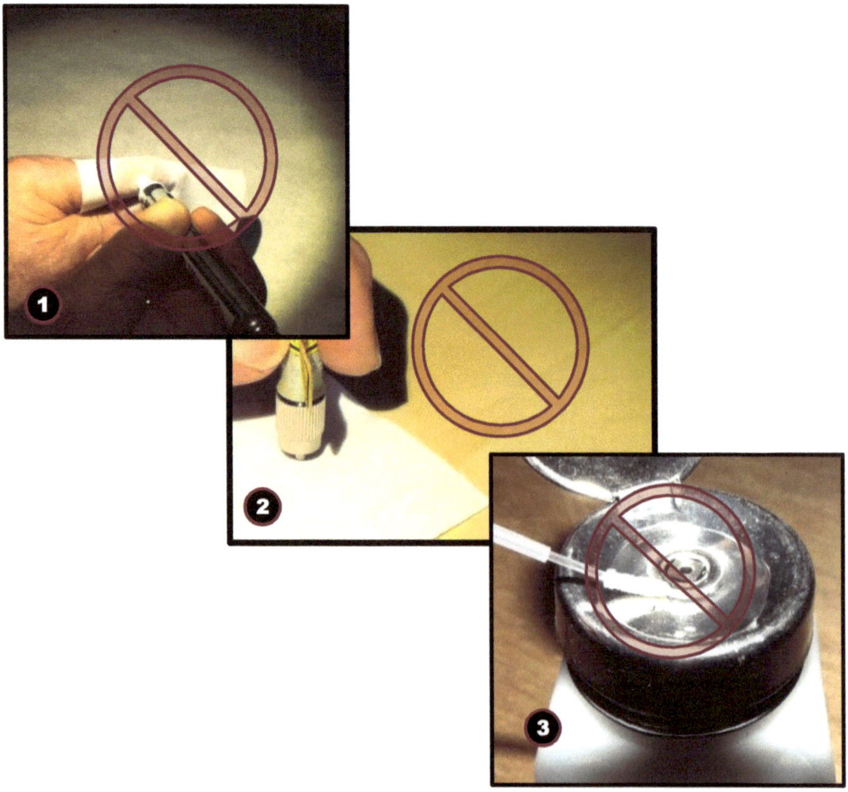

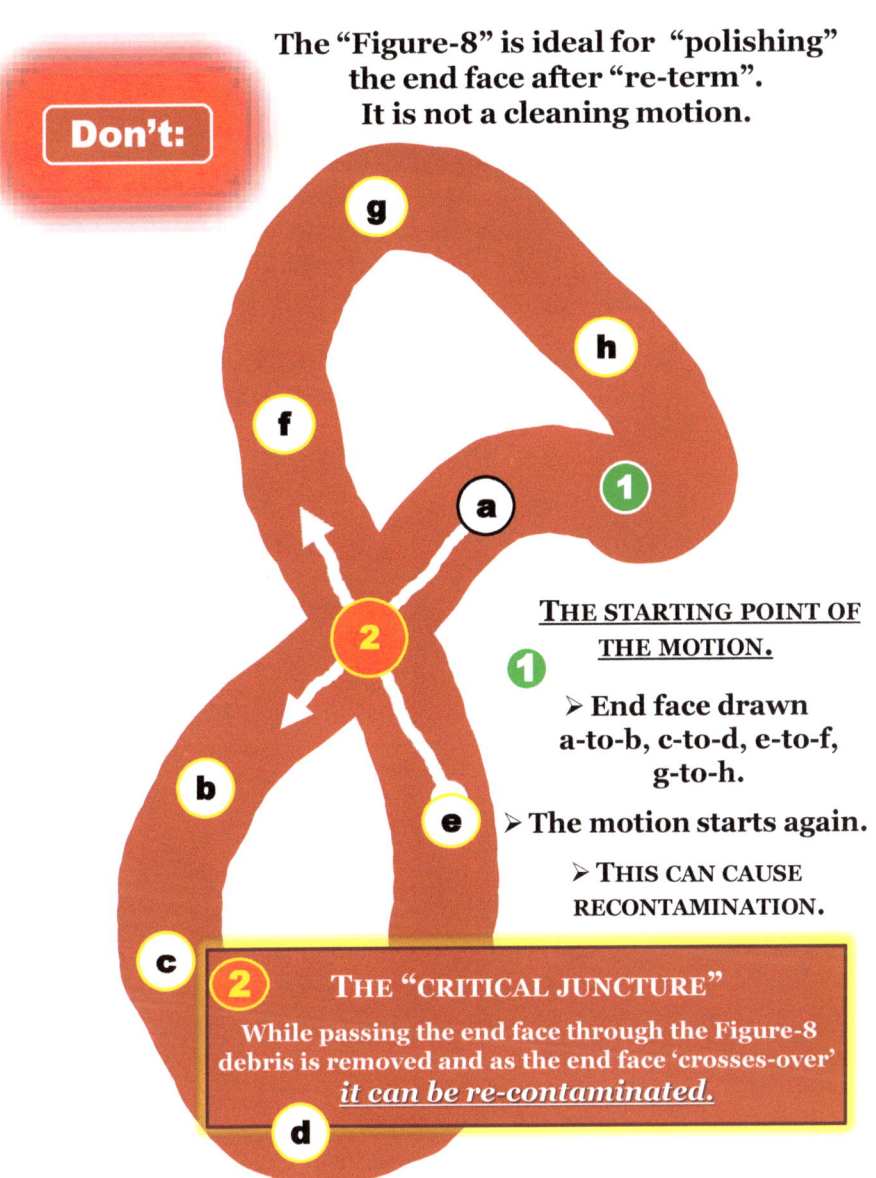

Conclusion

What Works!

This group of products always used with precision fiber optic solvents is "Best-Practice"

Eliminate the Confusion... change the procedure!

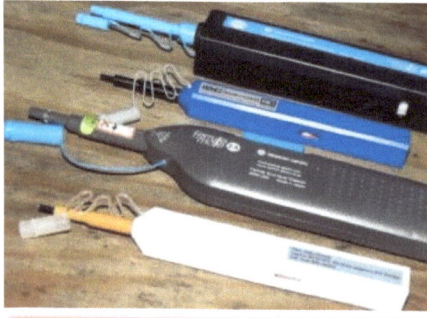

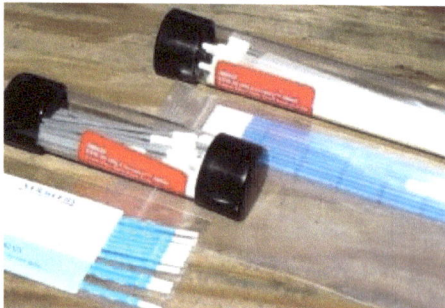

- ✓ **Which one?**
- ✓ It is an 'applications specific" decision.
- ✓ Coming up: check the comparison section: your selection and reflects the final result.

- ✓ Ask your rep!
- ✓ Request Samples & Training
- ✓ Challenge the factory!

Conclusion

Don't "dry clean" first ... always use a small amount of fiber optic grade cleaner with all tools.

THE FIRST TIME...EVERY TIME!

Eliminate the Confusion!

The Myths of Cleaning Products

Tips on How to Select Effective Precision Cleaning Products

The Myth of 99.9% "Reagent Grade" IPA

99.9% "Reagent Grade" Isopropanol is an effective fiber optic cleaning solvent

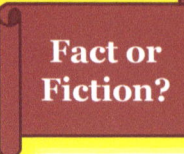

Fact or Fiction?

Two of the most successful replacement chemicals for 99.9% IPA are those based on 3M® HFE-7100 and some precision hydrocarbons. Remember, these are 'generic' terms: not all are the same. Test, study, sample against your anticipated contamination.

There are new solvents entering the market.
Personally, I think "aqueous cleaners" have promise as this chemical group is used on many electronics production lines and for precision metal cleaning.

The photographs in the following slides are from unbiased, vendor-neutral, formal laboratory testing.
There are other studies, please inquire.

(Special "thanks" to Chemtronics® for permission to use the images)

This testing used an aerosol container as the delivery system. Aerosols are clean and provide a constant flow. For testing of this type, delivery of the cleaning 'blast' was an important consideration.

Lower left lubricating grease is tested and lower right animal fat. When the procedure cleans "worst case" it's an indication of "best practice". You are encouraged to repeat this demonstration using contamination you encounter on the work site. Ask your manufacturer's rep for support.

The results are obvious: This particular precision hydrocarbon outperforms 99.9% Reagent Grade IPA.

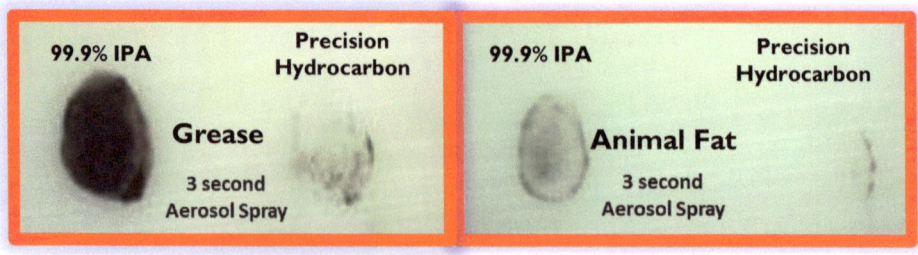

Please request this complete test panel as well as VideoLab® that compares fiberoptic cleaners against "worst case" contamination.

99.9% "Reagent Grade" Isopropanol is an effective fiber optic cleaning solvent

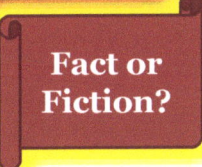

Fact or Fiction?

Some of the most popular fiber optic cleaners are based on 3M® HFE-7100. These products are packaged for several major producers by a well-established company. There are also aerosol versions of this chemical. One of the misconceptions of an aerosol container is that it has oily residue from the manufacturing process of the can. That perception is a marketing counter claim and a genuine "fiction'!

The real question is: "How do the HFE-7100 formulations compare to "precision hydrocarbons?"

The first thing to remember is that chemical companies carefully guard the proprietary nature of their formulations. Reading the SDS (Safety Data Sheets) provides insights into the contents. However, the proof lies in comparison testing such as this.

Testing is baselined using aerosol containers. The contamination was subjected to a 3 second aerosol spray.

Testing was 'vendor neutral".

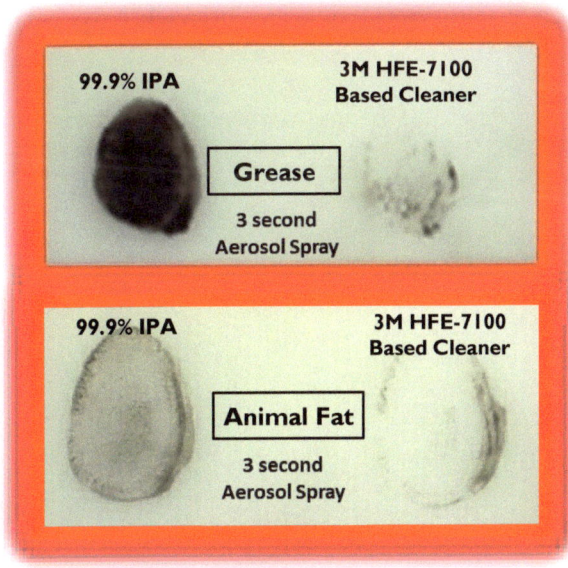

HFE-7100 outperforms 99.9& Reagent Grade IPA. These results are well known and accepted in other industrial segments.

WHY IS THIS SCIENTIFICALLY ACCURATE AND TRUE:

1. Isopropanol alcohol (IPA) is a 'polar solvent': it is effective cleaning 'polar soils' such as a finger print or 'salts'.

2. Precision Hydrocarbons, solvents based on 3M HFE-7100, DuPont® Vertrel® and some of the new Aqueous cleaners are designed to clean 'Non-Polar' soils such as pulling lubes, grease, hand lotions…the widest range you might imagine!

3. The "other problem" with IPA is that is it "hygroscopic"…<u>**it attracts moisture to itself rapidly, degrading an already weak cleaning solvent.**</u> Proper storage is an issue leading to degradation of the already weak cleaning chemical.

    - *If you can purchase 95-99.9% IPA in an aerosol container…it will stay fresher-longer. That said, it is still an ineffective and weak cleaner.*

BY THE WAY:
99.9% IPA is 'acceptable' for fusion splice prep ….
it's end face cleaning that suffers.

There are superior non-IPA cleaners for 'fusion splice prep' … that don't induce moisture into the splice *and possibly causing premature corrosion of splicer electrodes.*

<u>ALWAYS REMEMBER THE "ISSUE" OF IPA STORAGE.</u>

1. Solvent in the "well" absorbs moisture from the ambient.
2. Solvent "below the line" absorbs moisture from the air in the headroom "above the line".
3. Change IPA in these containers daily.

OTHER MYTHS OF CLEANING PRODUCTS

What else do you need to know about product selection?

What wiping material is better-best?

1.) **Clean room grade microfiber**
2.) **Hydro-entangled cellulose/polyester**
3.) ~~**100% paper and Low Lint Cotton**~~

"What did he just say, Vern?"

"He just said use just about anything except paper and cotton!"

It's helpful ... make that "important to learn" the jargon of your suppliers. Usually they will use 'speak in terms' as Vern and Jim are trying to understand. 'Learn the terms' and what they mean, in a practical sense, so you are on equal footing with those who want to sell you something!

Other "Myths of Cleaning"

"Beware of The Fuzzies"

Why is this important?

Remember, you are precision cleaning a precision surface that may be subject to lint. Challenge terms such as "low-lint". You are seeking as close to "lint-free" as possible. This exists in precision applications such as a wafer fab or medical operatory. Who is your seller? What is their experience in precision cleaning?

"Beware of The Fuzzies!"

There is no such thing as "lint free paper wipers".

Don't make this selection error!

Select "The Right Stuff"

"Lint-Free Wipers"
Marketing Claims and Reality

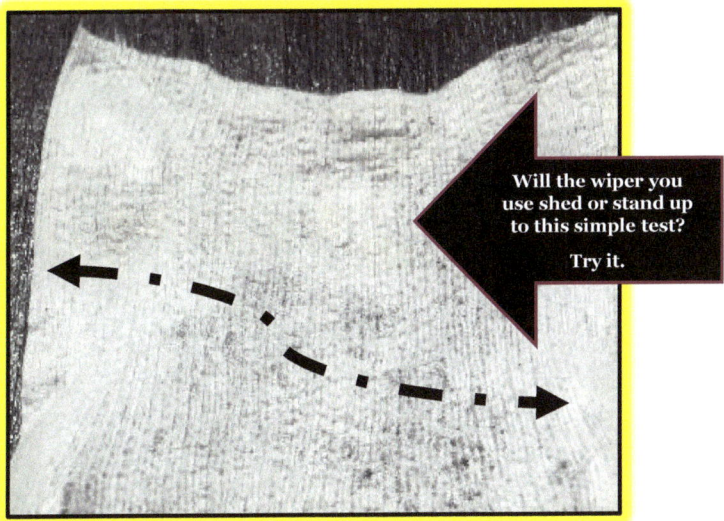

Will the wiper you use shed or stand up to this simple test?

Try it.

Note the elongation in this wiping material before it pulled apart.
Perform this simple "pull & tear test" before you buy a wiper.

Hydroentangled cellulose/polyester is strong and far more 'lint free' than 100% cotton or 100% paper.

Most probe tools and reel cleaners use a 'cleanroom grade' microfiber which is also exceptional.

Understand what you are being sold when you buy! Something as "simple" as a wiper can be the cause of unnecessary extra work...lost time. Unsatisfied client.

Other "Myths of Cleaning"

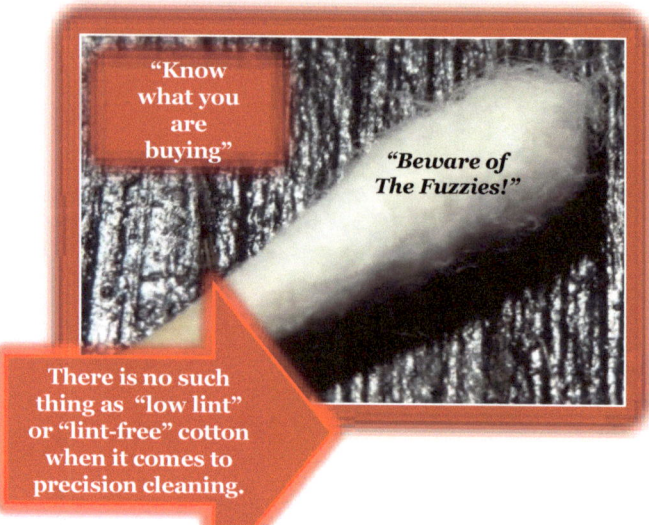

"Know what you are buying"

"*Beware of The Fuzzies!*"

There is no such thing as "low lint" or "lint-free" cotton when it comes to precision cleaning.

There are significant alternatives to "cotton".

One is medical grade 'foam' …and it's a myth that foam is not acceptable!

Others include clean room microfibers. Some of the most popular are 'sintered' or 'pull-truded' polyester.

Get the straight story from your suppliers! Sample and test before sending out a crew. Understand that your product selection has a direct impact on the result. You know this…and I believe that the wrong procedures have made precision cleaning fiber optic surfaces far more difficult than necessary.

It's for this reason that I encourage use of a fiber optic grade solvent with all products. This technique improves the way we clean just about anything. Time and again, since I began study of the topic, I have proven and re-proven to others…and myself…that product selection and how the product is used enhances fiber optic precision cleaning.

Would you think cleaning a fusion splicer with a "fuzzy" is a good idea?

Other "Myths of Cleaning"

"The Convenience of Probe Tools"

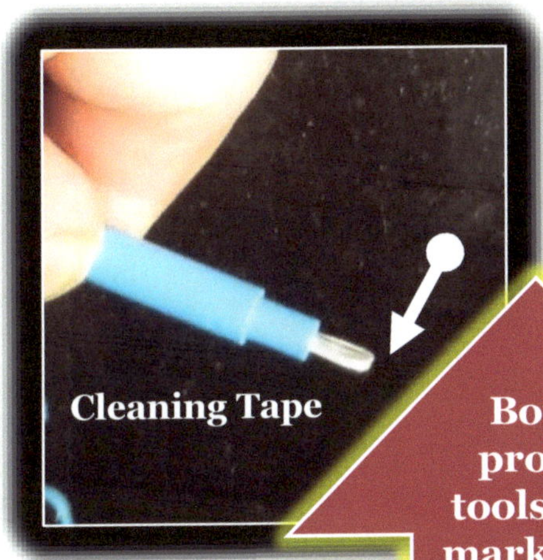

Cleaning Tape

Both probe tools are marketed as a 2.5mm cleaning device

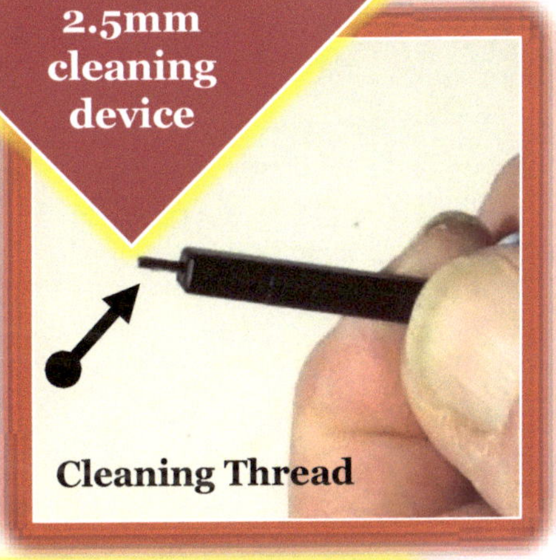

That's right...the white surface in the blue surround (top) and the white surface (barely seen) in the black surround (right) are the actual cleaning surfaces of these tools.

Which one will clean more of a 2.5mm surface?

Cleaning Thread

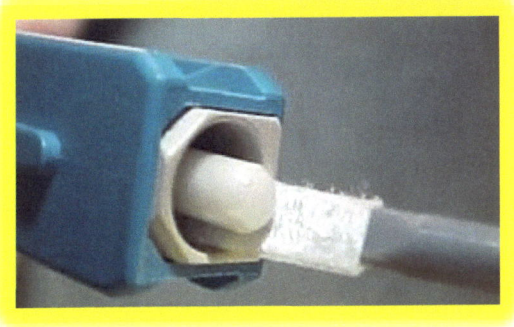

Precision swab tools are designed for applications specific cleaning.
Probe tools can't clean Zone-5 Vertical Surfaces

Remember this picture?

Only a well-designed, lightly moistened swab tool will clean the alignment sleeve. Appropriate fiber optic swab tools are constructed of microfiber, pull-truded or sintered polyester, hydroentangled poly/cellulose, or, foam. Some speak negatively about 'foam': remember, this material is used in medical operatories and Class-1 cleanrooms. Always ask if the foam is medical or cleanroom grade.

Images captured on RMS
Rotating Adapter Video
Inspection. (Patent Pending)

ATTENTION TO DETAIL:
CLEANING THE ALIGNMENT SLEEVE AND BACK PLAIN SURFACES

1.) This design as 'ridges' that ream the alignment sleeve in non-destructive cleaning. The tip is 'concave' to match APC and UPC surfaces.

2.) This "bullet tip" swab design cleans the back plane end face while the sides of the swab clean the alignment sleeve with a separate surface. The "bullet tip" folds over to clean UPC and APC.

3.) The 'hard flathead" pushes debris from the alignment sleeve forward to the back plane.

4.) The probe tip is small and has plastic sides (to assure the thread does not separate from the tool.)

Other "Myths of Cleaning"

"Optical" Pre-saturated Wipers
"Sure", they are convenient!

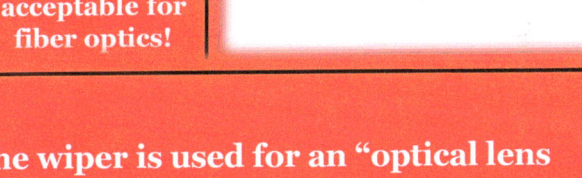

"...a *general all-purpose cleaner* is safe on plastics, PC Boards, connectors, fiber optics, semi-conductors, tape heads, office & medical equipment."

If the wiper is good for "everything" ... likely it's not acceptable for fiber optics!

If the wiper is used for an "optical lens surface", (microscopes, eyeglasses, camera lenses) *it's not acceptable for fiber optics!*

Some actually have surfactants: soap!

Eliminate "Myths of Cleaning"

"Fiber Optic Grade Cleaners"
Marketing Claims and Reality

Pens, Pumps, Aerosols

These are the "super stars" of the industry. Fiber Optic Grade Cleaners, along with those in other Industrial Segments, evolved beginning in 1999 with International Regulations eliminating CFC's.

Since that time, many industries have changed over. These are exceptional cleaners...worthy of our Industry.

When you select, always start with 'performance'. Don't be lulled into concerns about 'shipping' or packaging: make sure the products works for you...and you don't struggle to make it perform!

Check with the producers...understand "why" and "what" you are buying!

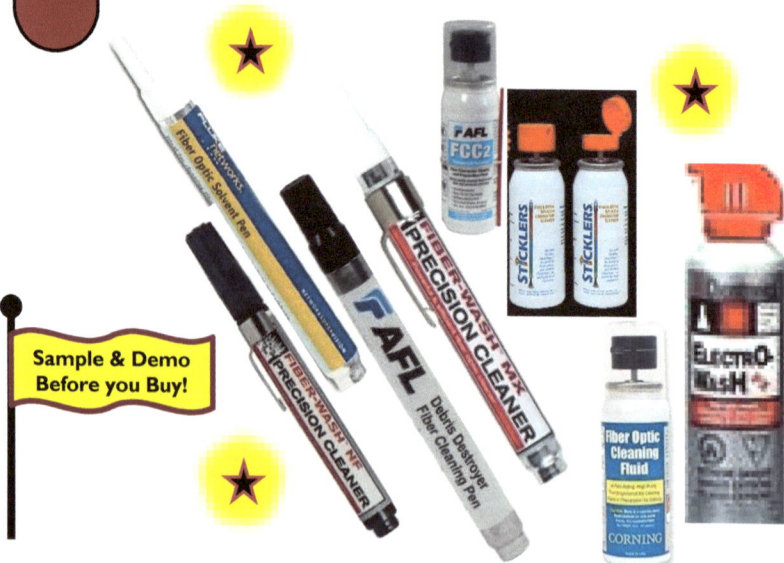

Sample & Demo Before you Buy!

Eliminate "Confusion of Cleaning"

"Audit". Is there another word that strikes more 'fear' in the hearts and minds of us all? !!! All this does is assure that chemicals are not used in the wrong places. I recall a visit to an airline technical operations center where a technician in the electronics lab borrowed a can of degreaser from his buddy in the landing gear area ... and destroyed a 1930's vintage Rockwell-Collins radio.

Know what the chemicals in your inventory do, can't do or will do if improperly used. Don't use that 'fear' as an excuse not to use contemporary chemicals! Learn the jargon of the trade, make your work easier and better.

ESTABLISH A "CHEMICAL AUDIT" PROGRAM

Contemporary chemicals are "applications specific"

- **Some industrial chemical formulations may NOT be appropriate for fiber optic applications**

  i.e.: Too strong and not 'plastic safe'
  Not strong enough and leave a "residue"

A SIMPLE AUDIT OF "WHAT IS ON SITE" IS ALWAYS "BEST PRACTICE"

	Chemical Trade Name	Department	Application	Notes
1.)	"FOC" Fiber Optic Cleaner	FTTx	Connector Cleaning	See "Red" Smith
2.)	"FSP" Fusion Splice Prep	FTTx	Fusion Splice	See: Julie Wilson
3.)	"GPC" General Purpose Cleaner	Maintenance	Wheel Bearings, Suspension	See: Rusty and Jill

Someone is thinking: *"This doesn't happen."* If it can...it will!

Other "Myths of Cleaning"

~~"Canned Air"~~

It is not *"breathable air"* ... it's a gas.

Read the label...some of these are "flammable gas"

I'll stop about ½ step short of using the term 'irresponsible' when describing compressed gas dusters as 'canned air'. There have been persons hurt who imagined these cans were breathable air...and maybe that will be enough to change the next ad or catalogue page.

What really matters is 'where do these chemicals fit' in the scheme of fiber optic cleaning and inspection?". Many years ago I was schooled about cleaning fusion splicers and while a one could imagine a compressed gas duster is ideal for this application...there is the possibility of launching a glass shard with the nozzle blast.

The nozzle blast might dislodge loose dust...and it also might excite dust around the connector area into a micro-sand storm! So, 'dust removal' doesn't seem like a good application for a compressed gas duster.

Having worked in the precision chemical industry for three decades, understanding how difficult it is for high performance degreasers to remove oily dust...it seems unlikely that a compressed gas duster will remove an oily residue.

In some instances, along with degreasing, came issues of static. It seems unlikely that a compressed gas duster will reduce or eliminate static field attraction of dust on a fiber optic connector...or anything for that matter!

Since most compressed gas dusters are refrigerant gasses, the one thing that most compressed gas duster can do it emit a -65F freezing coating to an end face. This is not desirable. That brings this point and that is: "If you must use a compressed gas duster...don't shake it."! Somehow in the liturgy of aerosols we shake before use! That's necessary for paints...not compressed gas dusters or aerosol cleaners.

So...does a compressed gas duster have any place in the fiber optic cleaning tool kit?

Eliminate "Myths of Cleaning"

"Compressed Gas Dusters"

Most Compressed Gas Dusters can emit a "freeze" spray. This was the original intent: *defective component isolation*

If you MUST use the product...select one with 'all-way valves' and high velocity.

While I am not a "fan" of this product to clean fiber optic surfaces, there is an application that 'calls out' for each warehouse to inventory the right version of the container.

Some producers have 're-valved' the container to where it actually has a little recoil! This is good because a little compressed gas duster can do a lot of good! Not only is the valve set to a higher nozzle volume, it is set to that it will not (easily) emit the freezing gas. That's a very good thing.

The right container has value to dry a storm-damage surface prior to precision cleaning as has been defined in this session. These are not 'precision cleaning'....

Conclusion

"Proper Product Selection assures Best Practice Results"

Fiber Optic Precision Cleaning and Inspection is an Applications-Specific Interactive Function.

There are actually two fiber optic cleaning procedures!

1. One is for end face and connector precision cleaning
2. The other is for fusion splice prep

It not an over-simplification to state: "Don't confuse the two procedures." Select products designed for each task.

Fiber Optic Precision Cleaning and Connector Inspection is an applications-specific function.

The only practical way to know if the fiber is clean for fusion splice prep is reflected in the loss measurement at the splice.

There are also cable cleaning products!

Conclusion

Inspection and Cleaning ... of anything ... is a well-established 'SCIENTIFIC ROAD MAP' that began 5,000 years ago with the invention of soap!

THIS 'MAP' is not often followed as fiber optic sciences evolved over the last 30+ years. You can update your work by creating your own internal standards for design, installation, and, training.

This is an important service to the Industry ...and your craft... until such time as Standards keep place with technology.

That also was our discussion.

Conclusion

- Cleaning a connector surface is not limited to the 'traditional end face'

- *There is more surface area to consider.*

- 'Soil points' at any location on connector surfaces may cause <u>primary</u> and/or <u>secondary</u> contamination

Images captured on RMS-1™ Rotating Adapter© Video Inspection. (Patent Pending)

Conclusion

→ If the fiber optic surfaces are not cleaned, and cleaned properly, there is opportunity for:

> Signal loss
> Inaccurate test
> A trouble call
> An unhappy client!

→ Inspection is critical and a video scope should be in your tool kit.

→ However, there are 'safety net' procedures...just in case!

→ Do not perform 'cross-over cleaning' by using products or procedures not intended for the task.

Conclusion

WHY WASTE TIME?

Why *"dry-clean first and then wet-to-dry clean if that doesn't work..."*

Always use a small amount of fiber optic grade precision cleaner...*with all tools* **The First Time...Every Time**

Conclusion

> "Best Process"
> is
> "Best Practice"!

"Thank you for purchasing this book and reading the story!

Questions, Challenges, or Comments?

My study, beginning in 1999, is based on meetings, formal visits, and interactions with you!

"The call is in your court!"

What did you see that made sense and what 'not so much'? With what do you agree...and... 'not so much'? Your input, as has been all along the way...will help all.

With my best to all...

Edward J. Forrest, Jr. "Ed"
+770-971-8100 (USA)
edforrest@fiberopticprecisioncleaning.com
edwforrest@gmail.com (Private)
www.fiberopticprecisioncleaning.com

Please, email your contact information so I might update you from time to time.

References

- Residual Contamination on Fiber Optic End Faces from Use of Polar Alcohols
  Paul Blair and Edward Forrest ITW Chemtronics June 2002 (rev:11/03) White Paper
- HFE-7100/Proprietary Formulation Cleaning Comparison to 99.9% Isopropanol
  James Fitzgerald. ITW Chemtronics   Research Chemist: 2003. Laboratory Test Soils a.) Animal Fat – Representative of skin and fingerprint oils Multipurpose Grease b.) LUBRIMATIC #11316 Motor Oil c.) Quaker State 10W 40 d.) Silicone Oil Dow Corning representative of pulling lube and buffer gel
- HFE-7100/Proprietary Formulation Cleaning Comparison to Precision Hydrocarbon/Proprietary Formulation
  James Fitzgerald. ITW Chemtronics Research Chemist: 2003. Laboratory Test Soils a.) Animal Fat – Representative of skin and fingerprint oils b.) Multipurpose Grease – LUBRIMATIC #113163) c.) Motor Oil Quaker State 10W 40 d.) Silicone Oil – Dow Corning representative of pulling lube and buffer gel
- A Study of 99.9% Isopropanol Absorption Rate of Water from Air
  Susan Max: Lead Chemist. ITW Chemtronics® 2004. Laboratory Test.
- A More Effective Means of Cleaning Fiber Optic Connections in FTTH, Outside Plant, and, OEM Applications. White Paper-Edward J. Forrest FTTH-2005
- Inspection and Cleaning Procedures for Fiber Optic Connections
  All contents © 1992–2006 Cisco Systems, Inc. Document ID: 51834 8-26-2006
- Soil Removal from End Face Utilizing Cisco Series of Ten Diverse Soils:
  Paul Blair, Ed Forrest, And Susan Max ITW Chemtronics: 2006. Laboratory Test
- Generating a Static Field When Precision Cleaning a Fiber Optic Connection:
  Paul Blair, Susan Max, and Edward Forrest. ITW Chemtronics: 2008 Lab Test
- TIA 455-240 September-2009
- IEC 61300-3-35 ed1.0 January 2015
- Telcordia GR-2923-CORE. February-2010
- Contemporary Considerations When Precision Cleaning Fiber Optic Connections: Performance-Inspection-Environmental Matters: White Paper Edward J. Forrest: November-2010
- SAE AIR 6031. 2012 Cleaning fiber optic connections.
- Comparisons of various cleaning solvents acting on ten complex soils and Investigation of Contamination of the Horizontal and Vertical Ferrule. Laboratory Test recorded on Video.   Edward J. Forrest: January 2011
- Interferometer readings courtesy of Promet Corporation using a FiBO® 250 device.
- Contemporary Considerations When Precision Cleaning a Fiber Optic Connection: 2011v8 Edward J. Forrest: Power Point. "Comparisons of cleaning techniques with audio and video".
- Bill Woodward: "FOI" Fiber Optic Installer. (ETA). Published by SYBEX
- "VideoLab" Tests of Various Cleaning Procedures on Simple and Complex Contaminants".
  Edward J. Forrest, Jr. August-2015
- "Clean and Inspect: What IEC 61300-3-35 means to you". -2015 Cabling and Installation. Brian Teague-MicroCare Corporation. 07-15"
- An Evaluation of Aqueous Cleaning Processes for Fiber Optic End Face Connections". VideoLab Study of commercial and experimental aqueous cleaners.   PowerPoint Edward J. Forrest, Jr. August-2015
- "The Need for Processes that Future Proof the Fiber Optic Installation"
  White Paper. Edward J. Forrest, Jr. July-2015. Updated July-2017
- "How we do and should not; should and may not, precision clean and inspect a fiber optic connection". Training with video. Edward J. Forrest, Jr. June 2016_2017
- "The Inextricable Interaction Between Precision Cleaning and Precision Inspection of a Fiber Optic Connector".   Edward J. Forrest, Jr. Training Video-2017. Text Book: CreateSpace
- EOS/ESD Association, Utica, NY. www.esda.org
- YouTube® Ed Forrest regarding Video Labs and Training
- www.fiberopticprecisioncleaning.com
- "Ozone Layer". WikiPedia. September-2017
- "Breaking Through Myth to Reality. A Future Proof View of Fiber Optic Inspection and Cleaning". Edward J. Forrest, Jr. Training Video-2018. Text Book: Create Space.

Acknowledgements The author wishes to express appreciation to numerous individuals who, for almost twenty years, have helped form these theses into practical applications and field-proven results. *This list includes, in no order of magnitude, or, with no impression one or the other agrees or disagrees with these applications and associated thesis:*

Appreciation is extended to: Paul Blair, Terry Dant, David Kuklinski, William Woodward, Eric Martini, Michael Hackert, Larry Johnson, Michael Schneider, Dr. Tatiana Berdinskikh, Jim Hayes, Dr. Osman Geblizlioglu, Don Stone, Mark Baranuk, Douglas Teller, Carl Walz, Gary Tyler, Roger Heydinger, Darrell Smith, Kenneth Putnam, Frank Giotto, James Henry, David Zika, Bob Menard, George Bell, Charles Mason, Anthony Lowe, John Cotterill, Vincent Racine, Michael Chilicki,, Curtis Hill, Paul Looney, Gary Tyler, Jim Drain, Keith Hayes, Yann LeLuyer, John Mazukoski, Laurence Wesson, Mel Lesperance, Richard Ednay, Kim Teesdale, Bill Johnston, Brian DiMarco, Kirk Donley, Dan Morris, Brian Teague, WP Beverly, Michael Yeilding, Earle Olson, *and thousands of technicians, dozens of field sales engineers who have input their applications and experiences though many contacts and practical associations.* Likewise, appreciation is extended to Verizon, AT&T, BICSI, CenturyLink, Comcast, FTTH Council, IMSA, IEEE-Aerospace, iNEMI, FOI, OFC/NFOEC, OSP-EXPO, SAE-Aerospace, and SCTE, for speaking and technical platforms to interact with industry professionals on all levels. *Appreciation to LEMO, Amphenol, TE, Radiall, and Neutrik for technical support and wizardry! Thank you, all. There are others I cannot acknowledge for reasons all will understand.*

Publications: Please inquire about MP4 Training Sessions that coordinate with many of these books. Discounts extended to trainers, universities. Distributor inquiries welcome.

- "Comparison Study of Precision Cleaning Methods for all Fiber Optic Connectors" amazon.com. Edward J. Forrest, Jr. 2014
- "How to Precision Clean All Fiber Optic Connections". amazon.com. Edward J. Forrest, Jr. 2015.
- "Maintaining a Fiber Optic Fusion Splicer". Amazon.com Edward J. Forrest, Jr. 2015"
- Understanding Cross-Contamination Points on Fiber Optic Test Equipment". amazon.com Edward J. Forrest, Jr. 2015
- "How We Do and Should Not, Should and May Not Clean a Fiber Optic Connection" amazon.com Edward J. Forrest, Jr. 2015
- "The Inextricable Interaction Between Precision Cleaning and Precision Inspection of a Fiber Optic Connector. Amazon.com Edward J Forrest, Jr. 2016 (MP4)
- "Breaking Through Myth to Reality. A Future-Proof View of Fiber Optic Precision Inspection and Cleaning." . Edward J. Forrest, Jr. amazon.com 2018 (MP4)

Patents
- Wrapped Foam Swab for Fiber Optic Cleaning: 6,393,651 (2002)
- Swab with Pull-Truded Fiber Tip for Fiber Optic Cleaning: 6,795,998 (2005)
- Fiber Optic Component Cleaning Device and Method: 6,865,770 (2005)
- Method of Removing Matrix From Fiber Optic Cable: 7,125,494 (2006)
- Premoistened Fiber Optic Component Cleaning Tool w/ Integrated Platen: 7,216,770 (2007)
- Premoistened Fiber Optic Cleaning Tool w/Integrated Platen: 7,303,069 (2007)
- Raised Platen for Fiber Optic Component Cleaning Device: 7,552,500 (2009)
- Fiber Optic Component Cleaning Device with Grooved Platen: 8,336,149 (2012)
- Compact Fiber Optic Component Cleaning Device and Method: 8,336,149 (2013)
- Oscillating Fiber Optic Cleaning Tool (Application) 2015
- Fiber Optic Video Inspection System and Method: 9,638,604 (2017)
- A Method for Direct View Digital Fiber Optic Inspection (Application) May-2017

	True	False	
1.)			There are three general types of debris: something "dry", something a "fluid" and those in "combination"
2.)			"Dry debris" is best removed using the "dry cleaning method"
3.)			"Fluidic debris" is best removed using the "wet-to-dry" method.
4.)			Existing standards consider fiber optic surfaces in three dimensions
5.)			Existing video inspection "sees" a limited area of the end face.
6.)			A fiber identifier can be manipulated to determine if the fiber is actually clean.
7.)			A "direct view" microscope, properly filtered, is safe to view an active laser.
8.)			Cleaning the first time is not likely even using the best cleaning products.
9.)			Cleaning is a 'product based decision' and has little to do with the actual cleaning procedure.
10.)			It's a good idea when "wet-to-dry cleaning" to wait 3-5 minutes and re-inspect to assure no Zone-5 flooding
11.)			Dry cleaning works best for contamination that is "wet"
12.)			Wet-to-dry cleaning works best for contamination that is "dry"
13.)			The "hybrid/combination" process cleans the widest range of debris when used with non-IPA cleaners
14.)			100% Paper wipers, when folded in half, are lint free.
15.)			Never use a foam swab tool, even if the manufacturer designates the material as 'medical grade'.
16.)			Euclid of Alexandria defined geometry in two dimensions and rigorously set standards endure to this time
17.)			Fiber optic cleaning is an applications specific task matched to debris type and location on connector surfaces
18.)			Fusion splice prep and end face cleaning are the same procedures
19.)			The "Figure-8" is a cleaning technique
20.)			99.9% "Reagent Grade" IPA is an effective cleaner on all types of debris.

	True	False	
1.)	X		As well, there are infinite subsets of these three types that make may make any cleaning process a challenge!
2.)		X	"Dry debris" is best removed using a moistened procedure with a true 'lint free' wiper for fiber optic aps.
3.)		X	"Fluidic debris" is best removed using a dry process. It's a "mopping action"..
4.)		X	Existing standards consider fiber optic surfaces and debris in two-dimensional diameter.
5.)	X		Most typically, a 400x video scope only considers a 250-300 micron radius of the actual end face.
6.)		X	"Seeing is believing". The only means to assure a fiber optic surface is clean is to use video inspection.
7.)		X	My personal opinion is to never take a chance with your eyesight.
8.)		X	The "Telcordia Process", as proven by clear testing, can approach 1st time cleaning at high likelihood of success.
9.)		X	Cleaning is an applications specific procedure.
10.)	X		If you "wet-to-dry" clean, it is possible to flood "Zone-5". Wait 3-5 minutes and re-inspect. The time "post cleaning" is as critical as time in test such as "pass-fail".
11.)	X		Dry cleaning works best for contamination that is "wet"
12.)	X		Wet-to-dry cleaning works best for "dry" contamination
13.)	X		The "hybrid/combination" process cleans the widest range of debris when used with non-IPA cleaners
14.)		X	100% Paper wipers, are never lint free.
15.)		X	There are thousands of foam types. If a producer chooses "foam", likely they have made a choice based on experience and performance. Challenge all producers.
16.)	X		Euclid of Alexandria defined geometry in two dimensions and rigorously set standards endure to this time
17.)	X		Fiber optic cleaning is an applications specific task matched to debris type and location on connector surfaces
18.)		X	Don't confuse the products and the procedures.
19.)		X	It's a polishing action.
20.)		X	99.9% "Reagent Grade" IPA only works on a limited soil range and is strongly influenced by moisture attracted to it in storage.

Additional Resources in this Series:

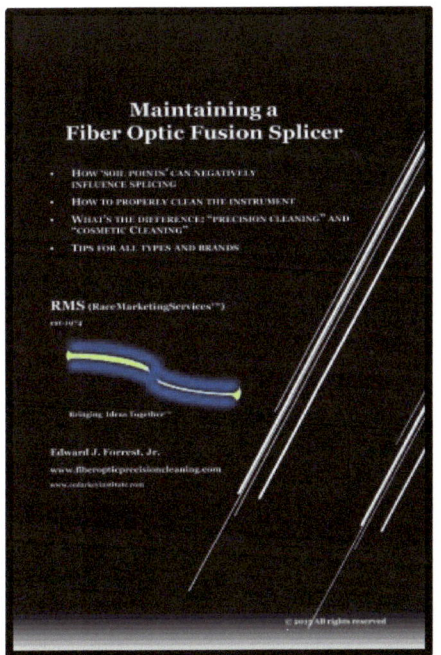

2014 Study updating "The Cisco® Series"

How to maintain a fusion splicer for lowest loss.

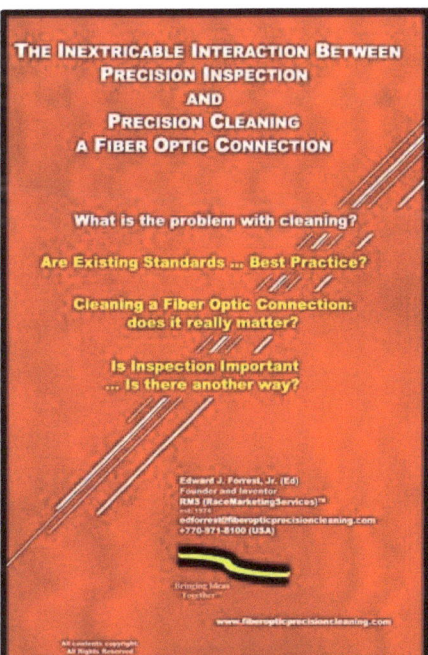

Short Course on Inspection and Cleaning

Available on line:

www.amazon/com (International)

Inquire about custom training programs
and product development support.
edforrest@fiberopticprecisioncleaning.com
+770-971-8100 USA

www.ingramcontent.com/pod-product-compliance
Lightning Source LLC
Chambersburg PA
CBHW040219220526

45473CB00001B/43